J. ROGER PITBLADO is a member of the Department of Geography at Laurentian University.

The failure of many of Tanzania's rural development schemes and policies results from their incompatability with existing agricultural systems. To date, information necessary for the assessment of such schemes has been sparse, particularly with respect to soils and land tenure, the agricultural subsystems frequently cited as sources of bottlenecks to rural expansion.

In this book Professor Pitblado describes and evaluates these subsystems and their possible influence on agricultural development of the North Mkata Plain. The study area (2740 sq. km in east central Tanzania) had previously been identified as an area for potential rural development by the International Bank for Reconstruction and Development.

In addition to presenting soil survey data recovered in the field, the author classifies and maps the soils by capability groupings indicating the degree and kind of limitation that would be faced in using each for general agriculture. Pitblado reviews relevant Tanzanian legislation for the period 1884-1978, revealing how the two major forms of rural tenancy — rights of occupancy by grant and by customary law — evolved. These modes of occupancy are examined at a reconnaissance level throughout the area and in detail for the section of the plain where land tenure rules were governed by the customary laws of the Ngulu people.

Information on the two subsystems is integrated by examining their influence on current land use and on future agricultural production. The author concludes that agricultural development is only ecologically feasible in selected areas through changes in the land use patterns and management practices in areas now under cultivation, and warns that some of the recommended changes will be inhibited by both historical and current land tenure rules and regulations.

University of Toronto
DEPARTMENT OF GEOGRAPHY
RESEARCH PUBLICATIONS

J. ROGER PITBLADO

The North Mkata Plain, Tanzania: A Study of Land Capability and Land Tenure

PUBLISHED FOR THE UNIVERSITY OF TORONTO
DEPARTMENT OF GEOGRAPHY
BY THE UNIVERSITY OF TORONTO PRESS
TORONTO BUFFALO LONDON

© University of Toronto Press 1981
Toronto Buffalo London
Printed in Canada
Reprinted in 2018

ISBN 978-0-8020-3378-9 (paper)

Canadian Cataloguing in Publication Data

Pitblado, J. Roger (John Roger), 1944-
    The North Mkata Plain, Tanzania

    (Research publications / University of Toronto. Dept. of
    Geography; 16)
    ISBN 978-0-8020-3378-9 (paper)
    1. North Mkata Plain (Tanzania). 2. Soils — Tanzania
    — North Mkata Plain. 3. Land tenure — Tanzania — North
    Mkata Plain. 4. Land use, Rural — Tanzania — North
    Mkata Plain. I. Title. II. Series: Research publications
    (University of Toronto. Dept: of Geography) ; 16.
    HD 987.Z63P57   333.76'09678'25   C81-094853-2

Preface

Many of Tanzania's rural development schemes and policies have been unsuccessful because they have been formulated and implemented without an adequate examination of whether they were appropriate to existing agricultural systems. Unfortunately, planners in Tanzania frequently find that there is little or no information available to them on which to base judgments that would be relevant to specific projects.

The information needs are particularly great with respect to the capability of soils and the characteristics of land tenure systems, two items often cited as sources of bottlenecks to Tanzanian rural development. The major aim of this study was to examine these items and evaluate their possible influence on the agricultural development of the North Mkata Plain, located in east central Tanzania. This location was selected for study because it had previously been identified as an area that might have potential for rural development.

The major field research for this study was carried out in 1970 and 1971, supported by the International Studies Programme of the University of Toronto. During that time I was also aided with a small travel grant as a research associate of the Bureau of Resource Assessment and Land Use Planning (BRALUP), University of Dar es Salaam. The initial findings of the study were reported in the form of a doctoral dissertation, accepted by the Department of Geography, University of Toronto, in 1975. Subsequent yearly visits to Tanzania on other projects in the period from 1975 to 1978 have enabled me to update some of that work. Funding for the latter projects was provided by the International Development Research Centre (Ottawa).

I have been assisted in the field work and presentation of this study by many individuals to whom I owe my appreciation. In Tanzania these included: Dr L. Berry, formerly the director of BRALUP; Dr A. Kesseba and Dr M. Meghji, formerly of the Faculty of Agriculture, Morogoro; Dr A. Uriyo, chairman of the

Department of Soil Chemistry, Morogoro; my field assistants, M. Shatner and J. Halfani; the librarians of the University of Dar es Salaam and the Dar es Salaam Public Library; the many Tanzanian officials who so willingly gave time and information; and particularly the Ngulu of the North Mkata Plain who generously accepted a stranger in their midst. In Canada technical assistance was provided by: the Extension Staff of the University of Guelph; Mr A. LeBlanc, a colleague formerly of Laurentian University; and especially Mr R. Labbé and Mrs M.C. Porter, technologist and secretary, respectively, in the Department of Geography, Laurentian University.

My sincere appreciation is also extended to the members of my thesis committee, especially Dr D.P. Kerr and Dr J.J. Van der Eyk. A very special debt of gratitude is owed to Dr Van der Eyk, my thesis advisor, for his patient advice and criticism from the formulation of the study through to its completion.

I would also like to thank Dr P.W. Porter of the University of Minnesota for providing me with potential evaporation data for several locations in the study area.

Finally, there is my wife, Jane. She has shared in both the joys and the difficulties of working in Tanzania. And she has struggled with the typing and editing of numerous drafts of this book. The work would not have been completed without her constant encouragement and assistance.

This book has been published with the help of a grant from the Social Science Federation of Canada, using funds provided by the Social Sciences and Humanities Research Council of Canada.

J.R.P.
Sudbury, Ontario
1981

Contents

viii Contents

1 At Magole the stream channel of the Mkundi River is sometimes completely filled in with coarse sediments derived from the eroded slopes of the Kidete and Nguru land units. These deposits reduce the capacity of the channel causing the river to flood frequently in periods of high water flow.

2 Approximately ten kilometres downstream from Magole the Mkundi River meanders across the Kilosa-Turiani land unit and is braided in some stretches.

3 The steep rocky slopes of Nguru ya Ndege support a vegetation cover of *Combretum-Brachystegia* woodland or miombo.

4 Fire-maintained grasslands and open wooded grasslands are the predominant vegetation types found in the Mkata Station–Dakawa land unit.

5 (upper) and 6   Deep wide cracking and surface mulches are common characteristics of the Mkata series. Gilgai microtopography occurs in localized areas. The major cracks in the soils of Illustration 6 are approximately five centimetres in width. The rule in Illustration 5 is marked off in ten-centimetre intervals.

7 Zebu cattle owned by the Baraguyu graze bushed grassland in the Mkata Station-Dakawa land unit.

8 Erosion channels developing in an overgrazed area close to the Wakwavi Settlement Scheme in the Mkata Station-Dakawa land unit.

9 Sisal bulbils newly planted and waiting to be planted, Kimamba.

10 Baraguyu boma.

12 Rice field, Kigugu.

PART I

Introduction

1

General Introduction and Summary

Large numbers of scientists from many disciplines have responded to the challenge provided by the problems of development. Their response can be attested to by the now voluminous and ever-growing literature on modernization and economic growth. Yet little of the research that is reported in this literature is of relevance to, or beneficial to, those most in need of it, the developing nations of the world. It has been stated that 'about ninety-eight per cent of all research and development expenditures in the world go toward the solution of problems concerning the wealthier countries' (in Ginsburg, 1973). In the same article, Ginsburg deplores the scandalously low priority that geographers have placed upon research on non-Western societies.

In response to the sentiments and observations expressed above, this study contributes to the filling of the gap in our geographical knowledge of one of the developing countries, Tanzania. Specifically it deals with the soils and land tenure in the North Mkata Plain, located in the east central part of that country.

Tanzania is desperately poor. It is listed in the statistics of the United Nations as one of the poorest nations of the world. To overcome the poverty, Tanzania has decided to plan its social and economic development. Most significantly, it has chosen *agriculture* as the priority vehicle for development.

Prior to and since achieving independence in 1961, Tanzania has experimented with numerous rural development strategies. However, most of these have been stories of failure. A variety of specific reasons for these failures have been offered (McKay, 1968; Newiger, 1968; and others). But common to all, and encompassing all of the specific reasons that have been cited, has been inadequate planning. Conspicuously, development schemes and policies have been formulated and implemented without an adequate evaluation of whether they were appropriate in the context of the contemporary agricultural systems of the country.

Unfortunately, planners in Tanzania do not have available to them data on which to base such evaluations. For example, soils are an obvious component of most agricultural systems. But there are few areas in Tanzania with anything but exploratory maps and/or descriptions of the soils. There are even fewer studies of the character of the soils in terms of their capability to support agriculture.

Similarly, land tenure is often cited as a major barrier to agricultural development in Tanzania. Yet there are very few studies available that can be used to demonstrate how the contemporary land tenure systems affect the use of agricultural land. And fewer still that pinpoint reasons why current tenure rules and regulations are a hindrance to agricultural development.

The present study sets out not only to describe the characteristics of the soils and land tenure systems of the North Mkata Plain, but also to assess their influence on possible future rural development.

The North Mkata Plain covers an area of some 2,740 square kilometres, centred around the grid reference 6° 40′s and 37° 27′E (see Figure 1). It is a part of the Wami River Basin. This basin has been referred to many times in planning and development documents as having potential for development. In 1961, when speaking of the potential for irrigation agriculture in Tanzania, the International Bank for Reconstruction and Development stated: 'Not much is known in detail of this large river basin ... there are indications that soil and topographical conditions would favour such development.' The first in the series of Tanzania's five-year plans for socio-economic development (Tanganyika, 1964) provided for major development of river valleys in the Wami, despite the lack of detailed data.

In the present work the soils and land tenure systems of the North Mkata Plain were studied separately in the field. They were then integrated by examining their separate and combined influence on current land use and by suggesting how they might affect future agricultural development. Accordingly, there are four parts to the presentation. The remainder of Part I deals with the methodologies of the survey and provides the environmental setting of the North Mkata Plain.

Part II provides an assessment of the soils of the study area. The soils of the North Mkata Plain were examined by conducting a land capability survey. This began with a soil survey in which representative soils were described and soil associations mapped. Each soil unit was placed in a capability group indicating various degrees and kinds of limitations that would be encountered in using the soil for general agriculture. A land capability map was prepared using the soil associations map as a base.

The preliminary step in the land tenure survey, described in Part III, was a review of relevant Tanzanian legislation over the period 1884 to 1978. This review revealed how the two present modes of rural tenancy, rights of

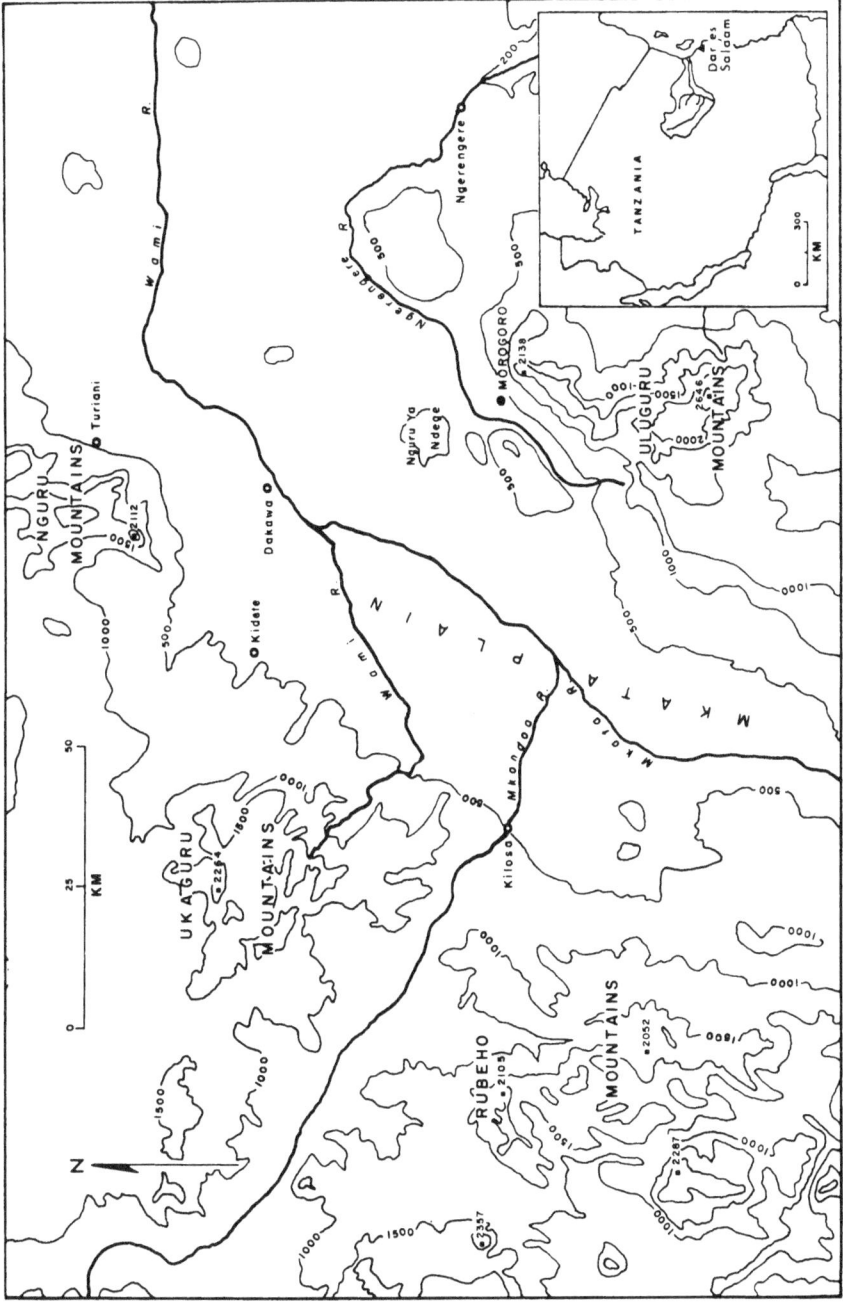

Figure 1 General location of the study area.

occupancy by grant and by customary law, have evolved. These modes of occupancy and their associated use of land were examined at a reconnaissance level throughout the study area. The character of the land tenure and land use systems was then analysed in a more detailed manner in a section of the North Mkata Plain where tenure rules are governed by the customary laws of the Ngulu people.

Part IV presents the author's final interpretations and major conclusions. It is shown that, given present levels of technology, the prospects for agricultural development in the North Mkata Plain through the expansion of the acreage under cultivation are severely limited. The majority of the soils of the region have physical and chemical characteristics that limit their capability to support arable agriculture. And in those areas where the soils are capable of supporting arable agriculture, the land is now either occupied or located where other factors limit its use. Some of these factors are the presence of tsetse fly, the absence of adequate supplies of moisture, or the fact that the land is held by a form of land tenure that prohibits expansion.

The study concludes that changes in land use patterns and management practices would make agricultural development of the North Mkata Plain ecologically feasible in the areas now under cultivation. Several examples of such changes are examined, but it is also suggested that some of the recommended changes will be inhibited by current land tenure rules and regulations.

2

Methods and Principles of Investigation

LAND CAPABILITY STUDY

The ability of a plant to grow, and especially to thrive, on a particular site depends on the ecological characteristics of that site and the adaptability of that plant to these characteristics. In areas of the world where experiments have been conducted, or where long-term yield records have been maintained, the ecological potential of a site to support agricultural crops can be stated with some degree of confidence. But in those areas where data of this sort are unavailable, especially a much too large section of the tropics, the potential of a site must be inferred. When the potential of a collection of sites must be assessed, it is common to systematize the inference process in what is known as a land capability survey.

Most land capability surveys are based on the concept that, for climatically adapted crops, the capability of land for agriculture is the result of a set of soil properties, a certain combination of which is considered to be optimal for crop production on a regional or national basis. At any one site, one or several or all of these properties may be less than optimal. Considering the properties separately and in combination allows one to judge the general potential for agriculture of that site in terms of the problems that may be encountered there, that is, the limitations of use. In a land capability survey, a site or a collection of sites is classified by placing it in that *class* (and *subclass*) that reflects its relative *degree* (and *kind*) of limitation to agricultural use.

In addition to placing sites into capability classes, most land capability surveys involve a mapping program. Units of land are delineated on a map and are labelled according to the dominating capability class, subclass, or some combination of these. In this way one is able to determine not only the location but also the spatial extent of various capability groupings within a study area.

The difficulties in such a survey arise from the complexities of the soil system and its relationship to various crops. Of the complex, interrelated factors of this system, which are to be described and evaluated? What assumptions are involved? What criteria are established for making particular or general inferences? What are the land units being mapped? And, in the final analysis, is the assessment reproducible or capable of being modified?

In responding to these questions, one must keep in mind that a capability assessment of land (whether for agriculture or for any other use) is an interpretive classification. The capability assessment therefore will depend not only on the classification itself but also on the nature of the available soils data. A proper understanding of a capability assessment is therefore enhanced when one is provided with the soils data on which the assessment is based.

Prior to the present study, the soils of the North Mkata Plain had been examined only at a very generalized, exploratory level. Milne devoted a few paragraphs to the soils of the area in a paper published (posthumously) in 1947; texture and pH values of five soil profiles were briefly described in 1957 by Sharma; a table of the analytical properties of one soil profile was published in 1969 by Lock; and in mapping the land resources of the southeastern quarter of Tanzania at a scale of 1:1,000,000, Johnson and Tiarks (1969) examined two soil profiles in detail in the North Mkata Plain.

Such limited soils data allow for only coarse, inter-regional capability assessments of the sort found in the *Atlas of Tanzania* (Tanzania, 1967a) — map scale of 1:3,000,000. As one of the major objectives of the present study was to provide for intra-regional capability assessments, much more information about soils was required. With so few previous data available, and constrained by limited financial and labour resources, the author found it necessary to conduct the capability study of the North Mkata Plain in three interrelated parts. The first part of the study required that the criteria for assessing land capability for agriculture be defined; the second involved a reconnaissance soil survey, with soils mapped at a scale of 1:250,000, thus providing the data that would be classified; and the third was the capability assessment itself. The procedures and principles involved in these interrelated aspects of the study are examined in the next two sections of this chapter.

Land Capability Classification for Agriculture
Using the foundations laid down by numerous authors of land capability classifications (for example, see Hardy, 1954; Klingebiel, 1958; Thomas and Vincent, 1959; Anon., 1965; or Obeng, 1968), the soils of the North Mkata Plain were assessed by a land capability scheme that was constructed with the following in mind:

1 It is assumed that only crops climatically adapted to the area under consideration will be grown.

2 Levels of technology beyond those currently employed in the survey area are not assumed. This assumption implies that soils are classified as they existed in the field at the time of the survey, and not as they might be upon reclamation or after other elaborate improvement projects.

3 Socio-economic factors, such as distance to market, market demand, land tenure, are not considered as criteria for capability groupings.

4 Although soil systems are dynamic, it is held that the relatively static characteristics of soil depth, natural drainage, texture, slope, and degree of erosion control, or at least influence, most of the more variable characteristics that together determine crop productivity. Therefore, it is assumed that these static, relatively easy-to-measure characteristics may serve as criteria to construct capability groupings. This assumption implies that in terms of available plant nutrients soil fertility is not directly involved in the construction of the capability groupings. It is assumed that, where the physical characteristics of a soil are ideal for crop production, productivity can only be improved by the proper application of manure or commercial fertilizer.

5 Given the above, it is assumed that fairly objective guidelines can be constructed which will allow capability groupings to be defined in terms of a combination of classes of the soil characteristics mentioned above.

6 The objective of the capability classification scheme is to identify and evaluate soils with respect to their general capability to support arable cultivation. Additionally, it is assumed that the morphological characteristics used in constructing the capability groupings will allow a measure of assessment of the capability of the soils for irrigation agriculture.

With the above comments in mind, a capability classification system was designed with three levels of generalization. The basic level is the *class*: the class gives the *degree* of limitation to arable use. A capability class can be subdivided into *subclasses*, a subclass indicating the *kind* of limitation. And finally, classes can be grouped into capability *divisions*, which allow for the mapping of land capability where the data do not justify the mapping of classes themselves.

It has been stated above that the capability groupings used to assess the land in the North Mkata Plain were constructed by examining a set of physical soil properties. A problem encountered here was the fact that prior to the survey it was impossible to obtain data that related crop productivity in the study area with these soil properties. Classes of these properties could not be established on the basis of information derived from the North Mkata Plain. It was necessary to choose soil property classes by adapting guidelines established elsewhere for capability classifications. Once these classes were chosen, and the guidelines for using them to construct capability classes were set, the guidelines were rigidly adhered to so that the classification would be reproducible and 'so that there is at least a well-defined basis on which changes can be superimposed later if and when they are felt to be justified' (Murdock and Andriesse, 1964). The six soil properties examined were depth, texture, natural drainage, degree of inundation, slope, and degree of erosion.

Depth
Depth was chosen as an assessment criterion because it is a measure of a soil's ability to provide a medium for root development and to supply moisture and nutrients to growing plants. Depth, in centimetres, was measured from the surface to 150 centimetres, or bedrock, or an impermeable layer, whichever came first. It was assumed that, other things being equal, the deeper a soil the greater would be its ability to provide root development and to store and supply moisture and plant nutrients. The ability of a soil to perform these functions decreases with decreasing depth. Furthermore, a shallow soil's ability to perform these functions would deteriorate more rapidly when subject to erosion than would that of a deep soil. Four depth classes were defined:

1 Deep – greater than 150 centimetres.
2 Moderately Deep – 90 to 150 centimetres.
3 Moderately Shallow – 50 to 90 centimetres.
4 Shallow – less than 50 centimetres.

Although the dividing lines between depth classes appear to be arbitrary, they are close to those selected elsewhere, which have been based on many years of testing and experimentation (Thomas and Vincent, 1959; Murdock and Andriesse, 1964).

Texture
While it has been stated above that 'other things being equal, the deeper a soil ... ,'' it must be recognized that one of the important soil properties that causes one to qualify the influence of depth is soil texture. Texture refers to the relative percentages of sand, silt, and clay found in a soil. Soils are considered coarse when they contain a high proportion of sand; they are fine when the clay fraction dominates; and they are medium-textured when the three soil fractions are present in relatively equal proportions. For most plants (there are exceptions) medium-textured soils are preferred.

A very coarse soil, while providing for ease in root penetration, is unable to store a great deal of moisture, whether it is deep or not. The other extreme, a very fine soil, decreases the ease with which roots may penetrate it but is able to store large amounts of moisture. In some instances, however, the moisture stored in fine soils is so great as to lead to the exclusion of air and hence to anaerobic conditions, which are unfavourable to plant growth.

Soil texture may also be used to draw inferences about a soil's ability to nourish plants. Whereas sands and silts are essentially unaltered, often inert primary minerals, most clays are composed of secondary minerals and are colloidal in character. The colloidal clay particles or micelles ordinarily carry a negative charge, attracting cations (plant nutrients such as potassium, magnesium, calcium) and holding them in exchangeable form for use by growing plants. In general, therefore, a greater proportion of clay in a soil will increase that soil's ability to store and supply plant nutrients.

Soil texture is determined in the field by feel and in the laboratory by conducting grain size analyses. Both of these procedures were employed in the present survey and are outlined in the following section under the title Soil Survey. The following criteria were used to define the four classes of soil texture employed in establishing the guidelines for constructing capability classes:

1 Very Coarse – sand, loamy sand.
2 Coarse – sandy loam
3 Medium – loam, silt loam, silt, clay loam, sandy clay loam, silty clay loam.
4 Fine – sandy clay, silty clay, clay.

Drainage and Degree of Inundation
In plants, water serves several functions: it is a necessary constituent of plant protoplasm; it is essential in the processes of photosynthesis and the conversion of starch to sugar; it enables plants to maintain turgidity; it is the solvent in which gases and materials move into plant roots and through the plants; and it is necessary for the process of transpiration by which plants are able to get rid of waste materials and are able to reduce excessively high temperatures (Russell, 1959). Both too little and too much soil moisture may have deleterious effects on plants. With too little water there is the danger that plants will be unable to obtain sufficient moisture to perform the functions listed above. With too much water there may be a significant reduction in soil aeration and, thus, in the levels of oxidation and solubility of several soil constituents. 'Because of the altered solubilities and ionic forms associated with anaerobic soil conditions, plants subjected to waterlogged conditions may show either toxicity or deficiency symptoms' (Russell, 1959).

In addition to directly influencing plant growth, water may also have an indirect influence on plant growth by affecting some physical soil properties. Water modifies the thermal properties of soils, the rate of movement of air into and through them, and their mechanical behaviour. It has been noted that 'dry soils are able to support heavy loads ... but added water reduces internal friction, making the soil susceptible to plastic deformation and consolidation' (Russell, 1959). Deformation and consolidation, evidenced by pans or puddled soils, may significantly reduce the effective depth of a soil and thereby limit its usefulness to plants.

The moisture regime of a soil can be measured in several ways. In the present study two field assessments were made. In the first, soils were placed into drainage classes that summarized the runoff, internal soil drainage, and permeability of the soils. The direct, quantitative measurement of these features is often difficult or at least very time-consuming and was not attempted in this survey. However, drainage classes can be inferred accurately enough for practical purposes by the close observation of indicator keys such as slope, soil colour, texture, structure, and consistency.

For most crops a well-drained soil is preferred. Soils in this drainage class normally occur on gently sloping to sloping land; soil colour is uniform throughout the soil profile and is often reddish or brownish in hue, indicating adequate oxidation; texture is in the medium range; soil structure is well-developed with the soil aggregates being small and subangular or rounded; and consistence can be described as non-sticky to slightly sticky and non-plastic to slightly plastic when wet, loose to friable when moist, and loose to slightly hard when dry. (Field determination procedures and specific definition of these features are detailed in later sections of this chapter.) Excessive drainage normally occurs when the soils are found on steeply sloping land. In this situation the soils will be coarse, structureless, and loose. Poor drainage is frequently indicated where the land is level and the soils are mottled, fine-textured, and poor in structure and consistence.

In the present survey five drainage classes were defined, which are indicated below using the descriptions that may be found in the *Guidelines for Soil Profile Description* (FAO, 1969):

1 Very Poorly Drained — Water leaves soil so slowly that the water table is at or above surface for most of the time. Usually occupy level or depressed sites; frequently ponded.

2 Poorly Drained — Soil remains wet for a large part of the time. Water table commonly at or near surface for considerable part of the year.

3 Imperfectly Drained – Water leaves soil slowly enough to keep it wet for significant periods but not all of the time. Often gley mottled in the lower A, or immediately below the A, horizon.
4 Well Drained – Water removed readily but not rapidly. May be gley mottled deep in the C horizon or below depths of a metre or so.
5 Excessively Drained – Water is removed from the soil very rapidly. May be steep, very porous or both. Profile free of gley mottling.

A second field assessment of the soil moisture regime used in constructing capability classes was the degree of inundation, which may be considered a special class of drainage. This special consideration was felt to be justified because of flooding problems in the North Mkata Plain. While flooding in itself may not be harmful – witness the dependence of agriculture on flooding in river basins such as the Nile and the Ganges – prolonged inundation frequently causes soil deterioration and crop damage and generally restricts agricultural use.

Three degrees of inundation were defined for this study on the basis of vegetation characteristics and information supplied by local farmers and engineers in the Morogoro office of the Tanzanian Water Development and Irrigation Department:

1 Nil – Soils that rarely experience flooding.
2 Moderate – Soils that normally experience annual inundation of short duration (less than 30 days). In uncultivated areas these soils commonly support a natural vegetation dominated by grasses, but bushes and trees are common.
3 Prolonged – Soils that are normally subjected to prolonged annual inundation. In uncultivated areas these soils commonly support a natural vegetation dominated by grasses. The presence of woody plants is rare.

Slope
Slope affects the capability of land for agriculture by influencing drainage characteristics, erosion potential, the use of machinery, and the costs of land preparation and maintenance. As the slope steepens drainage becomes increasingly excessive, the potential for soils to erode increases, the physical ability to employ machinery decreases while the operating costs increase, and the costs of land preparation and maintenance increase (Murdock and Andriesse, 1964).

Again, as data indicating the effects of slope are not available for the North Mkata Plain, the slope categories employed in this study are based on the intervals used elsewhere. Five slope classes were defined:

1 Flat — less than 2 per cent.
2 Gently Sloping — 2 to 5 per cent.
3 Sloping — 5 to 8 per cent.
4 Moderately Steep — 8 to 12 per cent.
5 Steep — greater than 12 per cent.

If other soil properties are favourable, surface irrigation would be possible on land in slope classes 1 and 2, although precautions to prevent soil erosion and the excessive drainage of irrigation water and leaching of soil nutrients would have to be taken. Surface, or more likely overhead, irrigation would be theoretically possible for the other slope classes but would require very major conservation works. With proper management, machinery could be employed on land in slope classes 1 to 3 but serious limitations would be experienced beyond these classes. Hand cultivation is also theoretically possible in all of these slope classes but is not recommended on land in slope class 5, and only with proper management practices in slope classes 3 and 4, because of the increasing hazard of erosion.

Degree of Erosion
While slope intervals may be used as a measure of erosion potential, another criterion for assessing the capability of land for sustained agriculture is the current state of erosion. Obviously the loss of soil through accelerated soil erosion reduces the land's capability for agriculture by the removal of topsoil — the soil's most valuable portion as a medium for root development and supplier of moisture and plant nutrients to crops. In addition, the costs of land use, care, and maintenance increase with increasing severity of erosion.

Relative degrees of erosion are probably best measured for agricultural purposes by noting the changes in productivity of eroding land over a period of time. As data of this sort are unavailable for the North Mkata Plain, a more subjective assessment was necessary. Through field observation and air photo analysis the areal extent of limiting erosion forms was estimated roughly. Five classes of erosion were then defined on the basis of proportions of land suffering from erosion. These classes, and the proportions used in their definition, are as follows:

1 Nil to Slight — less than 15 per cent.
2 Moderate — 15 to 30 per cent.
3 Slightly Severe — 30 to 45 per cent.
4 Severe — 45 to 60 per cent.
5 Very Severe — greater than 60 per cent.

Capability Classification

The land capability of an area is judged on the basis of the properties of representative soils found in the area (for a discussion of the latter aspect see the next section of this chapter, which deals with the soil survey). In the assessment of the agricultural capability of the land of the North Mkata Plain, five capability classes were defined. These classes are divisions of a graduated scale which indicates that the limitations to the general agricultural use of the land become progressively greater from class 1 to class 5.

Each capability class was defined in terms of permissible categories of the six soil properties described above. For capability class 1, for example, the permissible categories are: depth – deep; texture – medium; drainage – well-drained; inundation – nil; slope – flat to gently sloping; erosion – nil to slight. Therefore, for land to qualify as class 1, all six of these criteria must be met. If one or more of these criteria cannot be met, the land is downgraded to a class that incorporates the category of the property that is most limiting. The permissible categories of the properties used as classifying criteria are listed for each capability class in Table 1. Apart from being downgraded on the basis of one limiting characteristic, a soil can also be placed in a lower capability class because of the cumulative effect of two or more limitations of lesser magnitude.

Generalized descriptions of the five capability classes are given below:

Class 1 The soils of this class have very minor or no physical limitations to arable cultivation. The soils are deep, medium-textured, and well-drained and occur on flat to gentle slopes that suffer little from erosion. The land in this class is eminently suited for all forms of hand, mechanized, and irrigation agriculture.

Class 2 Moderate limitations reduce somewhat the value of these soils for arable cultivation. This class allows soils to be only moderately deep, fine-textured, and imperfectly drained. These soils may well provide adequate conditions for long-term agricultural use. Mechanized agriculture is not restricted, nor are irrigation practices, given proper management conditions.

Class 3 The limitations that relegate soils to this class are such that, while not excluding it, they will seriously restrict arable use unless specific management practices are employed. The use of mechanized techniques is equally advised only under strict management practices. Possibilities for irrigation agriculture are restricted.

Class 4 Land in this class is characterized by a combination of excessive slope and erosion, poor drainage and prolonged inundation, or other limiting soil

TABLE 1

Permissible categories* of the soil properties used in defining land capability classes

Capa- bility class	Depth	Texture	Drainage	Inun- dation	Slope	Erosion
1	1	3	4	1	1, 2	1
2	1, 2	3, 4	3, 4	1	1, 2	1, 2
3	1, 2, 3	2, 3, 4	2, 3, 4, 5	1, 2	1, 2, 3	1, 2, 3
4	1, 2, 3, 4	1, 2, 3, 4	1, 2, 3, 4, 5	1, 2, 3	1, 2, 3, 4	1, 2, 3, 4
5	1, 2, 3, 4	1, 2, 3, 4	1, 2, 3, 4, 5	1, 2, 3	1, 2, 3, 4, 5	1, 2, 3, 4, 5

*The categories, as defined in the text, are: *Depth*: 1, deep; 2, moderately deep; 3, moderately shallow; 4, shallow. *Texture*: 1, very coarse; 2, coarse; 3, medium; 4, fine.
Drainage: 1, very poorly drained; 2, poorly drained; 3, imperfectly drained; 4, well drained; 5, excessively drained. *Inundation*: 1, nil; 2, moderate; 3, prolonged. *Slope*: 1, flat; 2, gently sloping; 3, sloping; 4, moderately steep; 5, steep. *Erosion*: 1, nil to slight; 2, moderate; 3, slightly severe; 4, severe; 5, very severe.

properties such that even hand cultivation must be undertaken with caution. Non-arable practices, such as pastoral activities, are recommended.

Class 5 Slope and erosion features in particular exclude either arable or pastoral use of land in this class. This class of land would be best left for wildlife and/or forestry.

While five (or more) capability classes can readily be used in assessing representative soils, the mapping of the classes generally requires that the soils have been mapped in fairly great detail — at a scale of 1:63,360 or larger. As previously indicated, this study included a reconnaissance soil survey with mapping at a scale of 1:250,000. A survey of this type does not produce the amount or kind of data that would justify the mapping of land capability at the class level. By combining the classes, however, it is possible to provide a meaningful but generalized map of the land capability of the area.

In the present study three capability divisions have been defined by grouping the capability divisions as follows:

Capability division	Dominant capability classes
A	Classes 1 and 2
B	Classes 3 and 4
C	Class 5

Capability division A refers to land where more than 50 per cent of the soils would fall within classes 1 and 2 of the capability scheme. The areas so mapped are considered suited to sustaining most forms of agriculture with very few restrictions to arable cultivation. Land in capability division B, however, because the majority of the soils would be relegated to classes 3 and 4, has limited capability for agriculture; in many instances the land would be best suited to pastoral use. As the soils in capability division C would be assessed, for the most part, as class 5, the land so mapped should be restricted to wildlife preservation and/or forestry activities.

Capability divisions and classes indicate degrees of limitation but do not show the kinds of problems that would be encountered in using the land for agricultural purposes. To provide information on this aspect, subclass notations have been added to the division symbols (A, B, C) on the land capability map (Chapter 5). The notations used indicate the following factors as constituting the major limitations to agricultural land use: d, soil depth; t, soil texture; w, poor soil drainage; z, excessive soil drainage; e, soil erosion; s, steep slopes; i, prolonged annual flooding.

As an example, the notation *Bzt* would indicate that the area so mapped has limitations that restrict arable use but not pastoral use. The major factors that produce these limitations are excessive soil drainage and unfavourable texture. The order of the notations would indicate that excessive soil drainage is the more widespread of the two major limiting factors.

SOIL SURVEY

The determination of soil variations within an area is systematized in an inventory procedure known as a soil survey. The data derived from a soil survey are used to make deductions about soil formation and distribution and to make inferential assessments of the suitability or capability of land for various uses. In such surveys, soils are described, classified, and mapped in a documented, systematic manner.

Soil Description
Exposed to relatively constant environmental conditions for a considerable period of time, soils tend to acquire a layered arrangement of materials. The individual layers are called *horizons* and are best observed by examining vertical sections through the soils. The distinctions in the characteristics (i.e., the number, arrangement, and features of the horizons) of these vertical sections or *soil profiles* allow one to detect and define different types of soils. A soil profile is often referred to as being two-dimensional, but for its practical use a third

dimension is implied. This is a necessary addition because soils, being natural bodies, have width as well as length and depth.

A concept that makes the three dimensional character of soils more explicit is the *pedon* (Soil Survey Staff, 1960). A pedon is the smallest volume that can be considered to be a soil. It has three dimensions. More specifically, 'Its lower limit is the vague and somewhat arbitrary limit between soil and "not-soil". The lateral dimensions are large enough to permit study of the nature of any horizon present, for a horizon may be variable in thickness or even discontinuous. Its area ranges from 1 to 10 square metres, depending on the variability in the horizons ... Thus each pedon includes the range of horizons variability that occurs within these small areas ... The shape of the pedon is roughly hexagonal' (Soil Survey Staff, 1960).

In turn, the concept of the pedon is used to define the *soil individual*. Soil individuals consist of 'one or many contiguous pedons, bounded on all sides by "not-soil" or by pedons of unlike character' (Soil Survey Staff, 1960). A soil individual is distinguished from others by describing the features of one or more soil profiles that are representative of that individual's constituent pedons that have like character.

There is a vast array of soil profile features that can be employed to differentiate between pedons and between soil individuals. In practice it is not possible to examine all of these, but it is still possible to make meaningful distinctions on the basis of a small number of carefully selected profile characteristics. Unfortunately, in conducting soil surveys with the object of assessing land capability, the selection is sometimes too restricted. The data collected are often limited to only those soil characteristics used directly in constructing the capability classification. This is a wasteful practice, especially in developing countries where basic soil surveys may not have been undertaken. It means that the data collected are of limited use — relevant only to the goal of the specific survey itself and often not in a form that can be used in comparable areas.

This objection was kept in mind when choosing the characteristics that were used to describe the soils of the North Mkata Plain. These included both morphological and analytical properties. While some properties were not direct measurements of the capability classification criteria, they provided inferential data with respect to these criteria, as well as information that could be used to compare the soils of the North Mkata Plain with soils elsewhere. The morphological and analytical soil properties that were examined and that have not been fully discussed earlier in this chapter are listed below. Accompanying this list is a short explanation of how these features were determined and described.

Morphological Characteristics
Most of the soil properties listed here were determined from observations in soil pits. Seventy pits were excavated (see the section on Soil Mapping for how these pits were located) to a depth of 150 centimetres, to bedrock, or to an indurated layer, whichever came first. The major horizons of the exposed soils were determined on the basis of such features as colour, structure, texture, and root penetration. In some instances very thick uniform horizons were arbitrarily divided in half and each half described separately. Morphological characteristics were examined as follows:

1 Horizon Thickness – Measured in centimetres.
2 Colour – Determined by matching freshly broken pieces of soil with the colour chips of a *Munsell Soil Color Chart* (Munsell, 1954). Colours were commonly determined for both moist and dry states of the same piece of soil.
3 Mottling – Described in terms of abundance, size, contrast, sharpness of boundaries, and general colour. Terminology employed in describing these features is found in FAO (1969).
4 Texture – Determined in the field, by visual observation and feel, and in the laboratory. Procedures for the former determination may be found in FAO (1969) and, for the latter, in the section below headed Analytical Characteristics.
5 Structure – Determined by field observation and described in terms of grade (degree of aggregation), class (average size of individual aggregates), and type (form or shape of aggregates). Terminology and defintions of these expressions are found in FAO (1969).
6 Consistence – Resistance of soils to rupture or to deformation was determined in the field by feel and described under three soil-moisture conditions: wet, moist, and dry. Type of consistence and field determinations of them are given in FAO (1969).
7 Concretions – Described in terms of size (diameter in millimetres), type (iron, carbonate, etc.), and abundance (estimated percentage of the volume that they occupy in the horizon).
8 Roots – A general description of their size and abundance. Descriptive categories of root size and abundance may be found in FAO (1969).
9 Slope – Measured in per cent using a clinometer.

Analytical Characteristics
Soil samples from individual horizons of seventeen of the soil profiles described in the field were selected for laboratory analysis. The laboratory work outlined

below was carried out on material that had been air-dried and passed through a 2-millimetre screen. The analyses that were done are listed below:

1 Active Acidity (pH) – Determined using a glass electrode and an electronic pH meter. The soils were suspended in distilled water in a soil : water ratio of 1:5. The suspensions were stirred several times over a thirty-minute period, then the pH determined according to the method described by Peech (1965).

2 Organic Carbon – Determined by the Walkley and Black method using potassium dichromate as an oxidizing agent. This method is described by Allison (1965).

3 Total Nitrogen – Determined using the wet oxidation procedure of the Micro-Kjeldahl method described by Bremner (1965).

4 Total Phosphorus – Determined using the method of Saunders and Williams (1955).

5 Exchangeable Metal Cations – Soils were leached with a neutral ammonium acetate solution and the exchangeable cations – potassium, sodium, magnesium, calcium, and manganese – then determined spectroscopically.

6 Exchangeable Acidity – Determined by replacing the exchangeable hydrogen ions from the soil with barium using the Ba-acetate method outlined by Jacobs et al. (1971).

7 Total Cation Exchange Capacity (CEC) – Found by adding the sum of exchangeable metal cations to the amount of exchangeable hydrogen.

8 Base Saturation – Found by multiplying the sum of exchangeable cations by one hundred and then dividing by the total CEC.

9 Particle Size Distribution – Found by the pipette method. The soil fractions were expressed as percentages by weight of the following categories of particle size diameters: sand, 2.00–0.02 millimetres; silt, 0.02–0.002 millimetres; clay, less than 0.002 millimetres. These percentages were then used to determine soil textural classes by plotting the sand, silt, and clay percentages on a texture triangle.

10 Clay Mineralogy – In the absence of mineralogical data the shorthand method of Papadakis (1969) was employed to estimate whether the soils were dominated by 2:1 or 1:1 layer lattice clays. The procedure requires the calculation of an adjusted CEC:clay ratio. This ratio is equal to CEC/clay – CEC/100, where the cation exchange capacity is expressed in milliequivalents per 100 grams of soil and clay is expressed as a percentage. Accordingly, when the ratio is greater than 0.50, the soils are rich in 2:1 clays; when the ratio is between 0.50 and 0.15, the soils are rich in 1:1 clays but have a good deal of 2:1 clays; and when the ratio is less than 0.15, the clays of the soil are almost exclusively of the 1:1 type.

Soil Classification

To cope efficiently with a multitude of data one frequently reverts to classification. In soil studies this process has a number of objectives. Soil classification 'helps us to remember the significant characteristics of soils; to synthesize our knowledge about them; to see their relationship to one another and to their environments; and to develop predictions of their behaviour and responses to management and manipulation' (Kellogg, 1963) or 'helps in the generalization of basic data, in the understanding of relationships, in the conveyance of knowledge from one country to another, and in the conducting of adequate training, teaching, and technical assistance programs' (Van Wambeke, 1967).

These objectives of soil classification are commonly cited and agreed to by most scientists who are concerned with the character, genesis, distribution, and behaviour of soils. Consensus ends there, however. Because of the numerous philosophies of, and approaches to, soil taxonomy, soil classification, despite its rewards, is one of the most perplexing and frustrating aspects of soil study. Those classifications that are described as general purpose or natural are particularly confusing.

In contrast to special-purpose soil classifications (such as the land capability classification for agriculture that has been presented above), which are usually location-specific and designed to serve a utilitarian purpose, general-purpose classification systems are not location-specific, or are less so, and are designed for many uses (Avery, 1969; Macvicar, 1969). The ability of general soil classifications to serve many purposes derives from the fact that they group soils on the basis of their inherent characteristics, not inferred ones.

Numerous general classification schemes have been developed to group the soils of the world, a continent, a nation, or a region. But all of the schemes have met with some criticism, most of which centres around the problems of defining what criteria are to be used in the classification process (see, for example, Jones, 1959; and Webster, 1968). Despite the criticisms, many scientists, while accepting the present limitations, still feel that general-purpose soil classification is an important and necessary aid for soil studies. And many are making great efforts to improve on the existing systems.

One of the most important features of modern general soil classification schemes is the incorporation of the concept of the diagnostic horizon. 'The most fundamental and, possibly, the only real difference between soil and other unconsolidated geologic materials is that, in the case of soil, the materials have been organized by natural, non-depositional processes into horizons ... any soil can be described in general terms as consisting of a specific sequence of diagnostic horizons' (Van der Eyk et al., 1969). Thus, a number of diagnostic horizons can be defined in specific morphological and analytical terms, a soil can be described in terms of a specific sequence and number of diagnostic horizons, and taxa can be established in terms of permissible horizon arrangements.

Unfortunately, to date, no single general soil classification system has been established or received widespread support for use in Tanzania. No diagnostic horizons have been defined. And as the present survey was not designed to develop such a system, this approach could not be employed. Without defining diagnostic horizons, the soils of the North Mkata Plain were grouped on the basis of similar morphological properties, and the groups distributed among a single taxis, the *soil series*. This form of classification admittedly does not have the characteristics of general soil classification, but it does allow the possibility of correlating the soils of the study area with soils elsewhere because the soil series is the basic taxonomic unit of many general soil classification systems.

A soil series consists of 'a collection of soil individuals essentially uniform in differentiating characteristics and in arrangement of horizons' (Soil Survey Staff, 1960). And in many soil classifications the series is used to build up a hierarchy of soil categories or taxa. That is, a minimum of variation in soil properties is permitted when grouping soil individuals at the lowest-order taxonomic unit, the soil series. As one moves up through the hierarchy, the higher-order taxonomic units, consisting of an increasing number of soil series, allow for greater variations in soil properties at successive levels of abstraction or generalization.

While it has been stated above that no single general soil classification has been adopted in Tanzania, several systems have received a good deal of attention in recent years. One of these systems, *Soil Taxonomy* (Soil Survey Staff, 1975), uses the soil series as its basic taxonomic unit. An attempt was made to classify the soils of the North Mkata Plain using this system. The soils were also correlated with the SPI (Inter-African Pedological Service) system that was prepared for mapping the soils of Africa (D'Hoore, 1964). It should be noted that, while the SPI system is not strictly a general classification, 'the elements of the mapping units have, however, been grouped in such a way that the categories obtained can be compared with the higher categories of modern genetic classifications' (D'Hoore, 1968). A number of problems were encountered in using these classification systems to classify the soils of the North Mkata Plain, and these are discussed in Chapter 4.

Soil Mapping

For this study the soils of the North Mkata Plain were first mapped at the reconnaissance-level scale of 1:250,000 (Pitblado, 1975). The mapping procedure was aided by the knowledge, accumulated and recorded by many investigators over many years of study, that each different kind of soil does not occur randomly. There is a pattern of occurrence that, while often complex, can be deciphered. The pattern derives from the fact that the soils are the result of, and responsive to, the operation of five basic factors of soil formation: climate, vegetation, topography, parent material, and time. As these same factors in large measure control, or are features within, the landscape there is a close relationship between soils and landscapes.

The first step in mapping the soils of the North Mkata Plain was to subdivide the area into seven broad physiographic regions termed *land units*. The boundaries of a land unit would then encompass an area of relatively homogeneous climate, vegetation, topography, and parent material. If the premise that there is a close relationship between soils and landscapes was applicable to the study area, then each land unit should prove to be characterized by broadly similar soil characteristics. This was indeed the case, only minor boundary changes being required upon completion of the field work.

In fact land units of the type defined above had been delimited for the study area by Johnson and Tiarks (1969). However, as their map was of a scale of only 1:1,000,000, minor boundary changes were required for the present larger-scale survey. These changes were made primarily on the basis of sharp breaks of relief and associated vegetation differences. This task was accomplished after preliminary exploration of the study area by light aircraft and automobile and after examining all available topographic maps and air photographs.

The seven land units identified and mapped were then used in the next step of the soil mapping procedure. A decision was made to concentrate the major effort of the survey in those units that would likely be most promising for agricultural development. As four of the land units are dominated by very steep, rocky slopes, the survey of them was confined to the excavation of only a very few soil pits each. In each of the three remaining land units, a small key area that was representative of the environment of the land unit was selected for survey priority. The size, location, and characteristics of these land units and key areas are discussed in Chapter 3.

The key areas were chosen for the purpose of examining the range of characteristics of the soil individuals that would likely be found throughout the study area, defining the taxonomic soil classification units — the soil series — and discovering the geographical relationships between the soil series themselves and between the soil series and environmental features, such as topography and vegetation. The latter information would then be used as an aid to air photo interpretation and subsequent mapping of soil units.

Within each key area the boundaries of the soil individuals were traversed on foot, the boundaries being located by soil augering. The range of soil profile characteristics of the soil individuals was determined by excavating soil pits and describing the profiles in the manner outlined above. At the same time, observations were made with respect to land use, topography, surface drainage, and vegetation. In this part of the survey, 52 soil pits were excavated and approximately 2,000 sites examined by auger, giving an average observation density of 25 per square kilometre.

The information derived from these key area surveys was used, first, to define 23 soil series. Next, 14 soil *mapping units*, in this case *soil associations*, were defined generally as combinations of from two to four dominant series regularly associated in the landscape.

As defined, a soil association is a mapping unit and not identical with any taxonomic class of a general soil classification system. It comprises like and unlike soils (soil series) that occur together in systematic, geographically related patterns. In the present survey, as in other similar soil surveys, soil associations are defined in terms of the kinds and proportions of the like and unlike soils that comprise them. Usually several soil series can be observed as dominating a number of areas and occurring in relatively fixed proportions. Those areas are then mapped as belonging to a soil association having these particular characteristics. Other areas will be mapped as belonging to different soil associations depending on the kinds and proportions of the soil series that are present. For example, if two areas in question are made up of the same soil series but in differing proportions, then two associations will be mapped (Simonson, 1971).

It should be noted that the definitions of the soil associations in terms of their constituent soil series members and the air photo patterns used for their identification were modified where necessary by the reconnaissance examination of soils outside the key areas. In this part of the survey, the soils were examined on traverses made along all major thoroughfares and many minor ones, and along numerous footpaths, cut-lines, and game tracks, and sometimes in cross-country treks in a Land Rover. As road-cuts were few and far between, the soils were examined principally by auger, although this method was supplemented where possible with excavations of soil pits. When soil pits were dug, samples were collected from the described horizons. The number of soil pits and auger sites examined in this part of the soil survey, when added to those of the key areas, total 70 and 8,000, respectively. For the entire study area, this gives a site observation density of 3 per square kilometre.

In the present study, a soil association takes its name from one of the most widespread series included in the association. In this way, the dominant series are fully recognized and a general geographical location is indicated for the soil associations.

It should be recognized that inadvertent inclusions of soil series will be found within the boundaries delimited for an association, even though these soils have not been listed as belonging to that particular soil association. This lack of purity of the mapping units arises from the fact that, in a reconnaissance survey, the soils are mapped from limited field observations and by placing considerable reliance on air photo interpretation. Thus, a high degree of accuracy is sacrificed to decrease costs and to speed up the survey procedures. Although the degree of accuracy of the present survey cannot be stated precisely, the density of sites examined falls just short of the range (between 4 and 25 field observations per square kilometre) recommended by Young (1973). It may be noted that this difficulty is not confined solely to reconnaissance mapping – it is a problem encountered in all surveys at all mapping scales. For instance, Young (1973) states that 'the purity of mapped Soil Series, that is, the percentage of the mapped area that actually belongs to the Series indicated on the map, formerly assumed to be about 85 per cent, is usually found to be only 50–65 per cent."

In the field mapping of the soil associations, mapping unit boundaries were plotted on the basis of stereoscopic examination of air photographs. This process was controlled by the information gathered in the key areas and along the soil survey traverses. Thus, the exact placement of the soil boundaries relied very much on the recognition of vegetation, relief, and drainage photopatterns that could be related to soil variations.

LAND TENURE STUDY

The examination of the forms and character of agricultural land tenure and associated land use patterns in the North Mkata Plain involved the reviewing of directly relevant literature, the interviewing of selected individuals in local government offices and instructors in land law, exploratory field work throughout the study area and the analysis of air photographs, and a detailed field examination of a selected area. Unlike the procedures employed in the study of soils and land capability, many of which are common to all soil surveys, there are no standardized methods laid down to guide the step-by-step procedures of this part of the study. The activities that have been listed and that are outlined below did not necessarily occur in the sequential order in which they are presented here.

Literature Review
In this stage of the investigation, conducted primarily in the Library of the University of Dar es Salaam and the Dar es Salaam Public Library, an attempt was made to discover and examine all (or any!) published material relating to both land use and land tenure in the North Mkata Plain. Naturally, the object of this search was to find material upon which the later field investigations could be based.

The people of the North Mkata Plain are of course subject to the laws of the country, and the most profitable part of this stage of the investigation was the time spent reviewing the development of Tanzanian legislation pertaining to land tenure and land use. In this review, most of the legislative records themselves were examined as well as a number of relevant commentaries on them. A preliminary report of this work was prepared (Pitblado, 1970). A later publication by James (1971) has since been found to be extremely useful in looking at this legislation from a lawyer's point of view.

Selected Interviews

In order to gain an appreciation of the views held by Tanzanians responsible for administering or teaching the laws pertaining to land tenure and land use in the North Mkata Plain, interviews were held with a number of government officials and with instructors of land law at the Institute for Development Management, Mzumbe (Morogoro). In total, six interviews were undertaken as relatively informal conversations. However, the discussions were structured in the sense that a number of written questions had been provided one or two days in advance of the interviews. The questions, restricted to customary land rights, included:

A Methods of Acquiring and Transferring Land
If by allocation:
1 Who is the allocating authority?
2 Is a cash fee paid for this service? a gift?
3 Who is the recipient of the fee or gift, if any?
4 What rights does the person have who received land by allocation? For example, does the person have the right simply to use the land in his lifetime? Or, in addition, can he sell, rent, or lend land at will, and pass it on to his heirs as he sees fit?
5 Can an allocation of land be made to a stranger? If so, are there any special conditions under which the allocation is made?
6 Is the permission of an authority required for the use of uncleared land?
7 Is the permission of an authority required for the use of cleared but currently unused land?
8 Are boundaries of a holding marked?
9 Is there a limitation of the size of the holding a person may acquire?

If by inheritance (in addition to questions similar to those above):
1 Is the consent of an authority needed before inheriting a piece of land?
2 How is a holding divided among heirs?

3 Can females inherit land?
4 Can a person inherit land from more than one person?

If by purchase (in addition to questions similar to those above):
1 Can all types of land be sold?
2 If a person wishes to sell his land, do his immediate relatives have the right to make the first bid?
3 Are complete holdings sold, or only single fields?
4 If land cannot be sold, are some other entities? clearing costs? costs of improvements? trees?

If by loan or rent (in addition to questions similar to those above):
1 Are complete holdings loaned or rented, or only single fields?
2 Is rent paid? in cash? other forms of payment?
3 Can a person who loaned or rented a piece of land be evicted at any time? If so, by whom?

B Miscellaneous
1 To what extent have the establishment of local governments and the influence of agricultural field officers usurped the land allocating powers of the tribal elders?
2 For example, a man wants a new *shamba* (farm). The *jumbe* (chief tribal elder) shows him a piece of land but he is not satisfied with it for some reason. Must he accept this piece of land on penalty of not receiving any land at all, or can he argue the point with the jumbe?
3 'Development conditions' apply to rights of occupancy by grant. Do these also apply, either formally or informally, to rights of occupancy arising out of customary law?
4 In an area where there are marked differences in the quality of soil, does each member of an area have the right to a portion of the best land?
5 If a man wants to make improvements on his holding but does not have the finances available to do so, how can he obtain the necessary funds?

Despite the obvious willingness of the interviewees to assist in the survey, this part of the investigation produced little concrete information concerning the customary land tenure forms in the North Mkata Plain. The principal reasons were that the interviewees were either young, newly appointed, and inexperienced or could trace their own affiliation to tribes not found in the study area. In one interview, for example, more was learned about the Mwera of Kilwa District (southeastern Tanzania) than about any of the people of the study area.

Reconnaissance Survey

While conducting the reconnaissance soil survey of the study area, the author made general land use observations along the traverses and at each of the sites where soils were examined. It became quite apparent that the land use patterns associated with the two major divisions of rural land-holding could be distinguished very easily on the ground and through air photo interpretation using a simple classification.

Defined according to the dominant rural activity that was being practised at the time of the survey, the individual land use types, which are described in Chapter 6, are self-explanatory for the most part. They include:

I *Rights of Occupancy by Grant*
Sisal estates
Wami Irrigation Scheme
Wami Prison Farm
Nguru Forest Reserve
National Development Corporation Cattle Ranch
Ujamaa villages
Other largeholdings

II *Customary Rights of Occupancy*
Recently cultivated smallholdings
Unoccupied or periodically grazed land
Wakwavi Settlement Scheme

Detailed Survey

While the preceding stages of the study provided land tenure and land use information of a general nature, more detailed data from areas held by customary law were considered necessary. To obtain them, a small area was selected for intensive investigation.

In selecting this area, the author was aware that the North Mkata Plain is part of the territories traditionally occupied by two tribal groups, the Kaguru and the Ngulu. Indeed the study area straddles the ill-defined, diffuse boundary that separates the Kaguru to the south from the Ngulu to the north. While a fair amount of information on the agricultural systems of the Kaguru is available, little has been provided for those of the Ngulu. A decision was made then to select the area of detailed study from a location occupied predominantly by the latter group.

When exploratory work was being conducted for the purposes of selecting key areas for the land capability study, the above considerations were kept in mind. Therefore, one of the key areas was chosen not only for its location with respect to one of the land units but also for its location inside the Ngulu tribal territory. This aspect of the selection process was aided by referring to the tribal maps published by Gulliver (1959) and Beidelman (1967) and by consultations with several inhabitants of the study area.

The key area that was chosen is indicated in Figure 3 (Chapter 3), labelled Makuyu. The survey focused on two groups of villages in this area. For convenience, the village groups were named after the largest and/or oldest member villages. The locations of the villages within the Makuyu key area are indicated in Figure 2 and are listed below:

Makuyu villages	Kigugu villages
Makuyu	Kigugu
Mkololoni	Madegho
Kipinde	Makuto
Mzizima	Dibamba
Milongwe	
Muibuka	
Chamkole	

The land tenure-land use survey of these villages was conducted in two stages. In the first, conversations were held with village elders. Again, these conversations were conducted informally although guided by questions similar to those cited earlier in this chapter.

The second stage of this survey was concerned with the gathering of specific land tenure and land use data, collected by administering a questionnaire. It was intended that all heads of households in the eleven villages be interviewed. This aim was accomplished for only four villages — Makuyu, Mkololoni, Muibuka, Kigugu. Failure to survey the remaining villages in a like manner was due to the fact that adequately trained assistants were not available for a sufficient period of time.

Two hundred and sixty-five questionnaires were administered by Tanzanian assistants. Approximately 80 per cent of the interviews were conducted by one assistant in the presence of, and with the aid of, the local agricultural extension officer (*bwana shamba*). The remaining 20 per cent were conducted by this assistant and another. The author was present for only a small proportion of these interviews at the beginning of the survey. For illustration purposes, the land use in the area around the Makuyu, Mkololoni, and Muibuka villages was mapped by identifying the farm plots examined in the field and on air photographs. The area selected for this detailed mapping is shown in Figure 2.

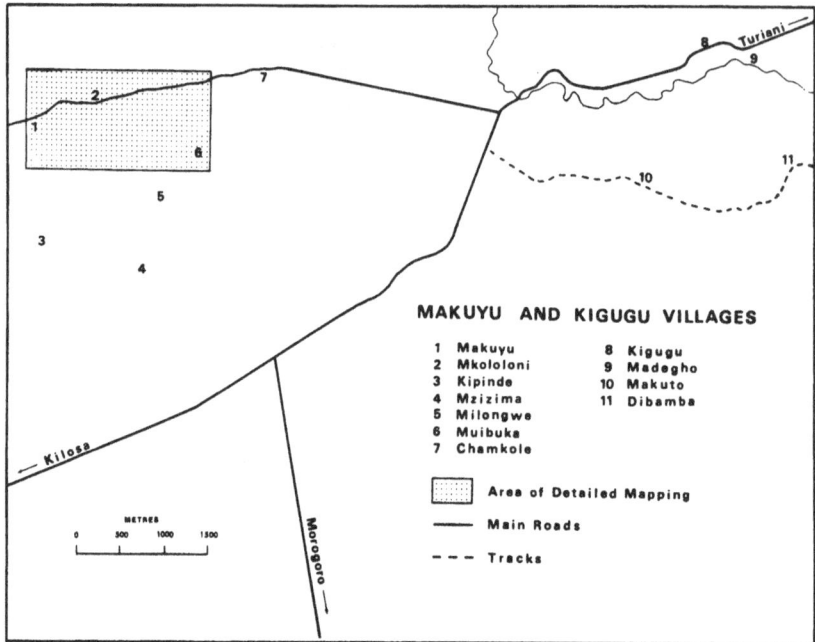

Figure 2 Location of the Makuyu and Kigugu villages in the Makuyu key area.

The questionnaire included the following items:

1 Size of Family – The household members were enumerated, a division being made between adults and children. Children were considered to be persons 15 years of age or younger.

2 Crop Types – For each field a farmer was asked to identify the crop or crops that he had grown there in the previous growing season. In some instances a check was made by examining the refuse in the fields, although it was found to be unnecessary as the farmers had no difficulty in supplying the information.

3 Crop Yields – Yields were estimated by farmer and/or the *bwana shamba* in terms of the number of *debes* (5 debes = 1 bag = 122 pounds) per field or per acre, whichever was the easiest or known value. (Note: It is only in the last few years that Tanzania has been converting to the metric system. Therefore, the British units of measurement, more familiar to the *bwana shamba*, the Ngulu farmers, and the field assistants, were used in the field and converted to metric units later.)

4 Livestock – The type of, number of, and purpose for keeping livestock were enumerated.

5 Cultivating Practices — For each of the activities of field preparation, sowing, weeding, and harvesting, the tools used and the time of year (month or months) in which the activities were conducted were identified. The amount of time (number of days) required to carry out the activities was estimated.

6 Marketing — Farmers were asked to identify the amount and type of crops sold and the location of their market or markets.

7 Occupancy — The length of occupancy and the method of acquiring each field were identified.

8 Soil Assessment — The farmers were asked to qualitatively assess the soils of their fields and to indicate how much longer they planned to cultivate them.

9 Farm Size — For each plot or field, the length of plot sides was determined by pacing and the angles between the sides measured using a hand compass. The size of each plot was determined by drawing the plot shapes on graph paper and then counting squares.

10 Distance from Plot to Domicile — This distance was measured from the centre of the plot to the centre of the village in which the interviewee lived. In one measurement, the distance was determined by pacing along the path or paths identified by the farmer as his most common route of travel. A second measurement, the straight-line distance between plot and village, was made using the air photographs.

Outside the key area the same questionnaire was administered at another 37 locations, in order to provide a limited check on the data gathered in the areas of the Makuyu and Kigugu villages. In total, then, 302 questionnaires were administered both inside and outside the key areas.

3

Preliminary Environmental Considerations

But for a few descriptive lines found in general references, no analyses are available of the geomorphology, climate, or vegetation of the North Mkata Plain. And not many more than a few paragraphs can be located that mention the geological and river drainage characteristics of the area. As these are factors that influence soil variability and land capability, an attempt has been made to piece together a description of the physical environment of the study area from these few lines and paragraphs and from field observations.

PHYSIOGRAPHY AND GEOLOGY

The central portion of the Wami River Basin has been subjected to block faulting in several erogenic periods during the late Tertiary, the Pleistocene, and as recently as the Holocene (Pallister, 1971). Consequently, two major relief features dominate the landscape today – lowlands and highlands. The lowlands comprise a north-south trending trough partially filled with alluvium and colluvium derived from the highlands which flank both its eastern and western margins.

As a result of their being highly dissected by erosional processes, the uplifted highland portions of the landscape have complex slopes and elevations ranging from approximately 500 metres to over 2,000 metres above mean sea level (AMSL). In contrast, depositional processes are much in evidence in the lowlands. Here the land is little dissected and relatively level, and elevations range only from 370 metres to 500 metres AMSL. The boundary between these contrasting landscape features, located at an elevation of approximately 500 metres, is distinct. In particular, it is here that the lowland slopes of less than 8 per cent break sharply with the slopes of the highlands. Changes in vegetation and climate occur here as well, but the distinctions are less abrupt.

With respect to the Wami River Basin, the lowlands are generally referred to collectively as the Mkata Plain or the Central Alluvial Plain. The highlands com-

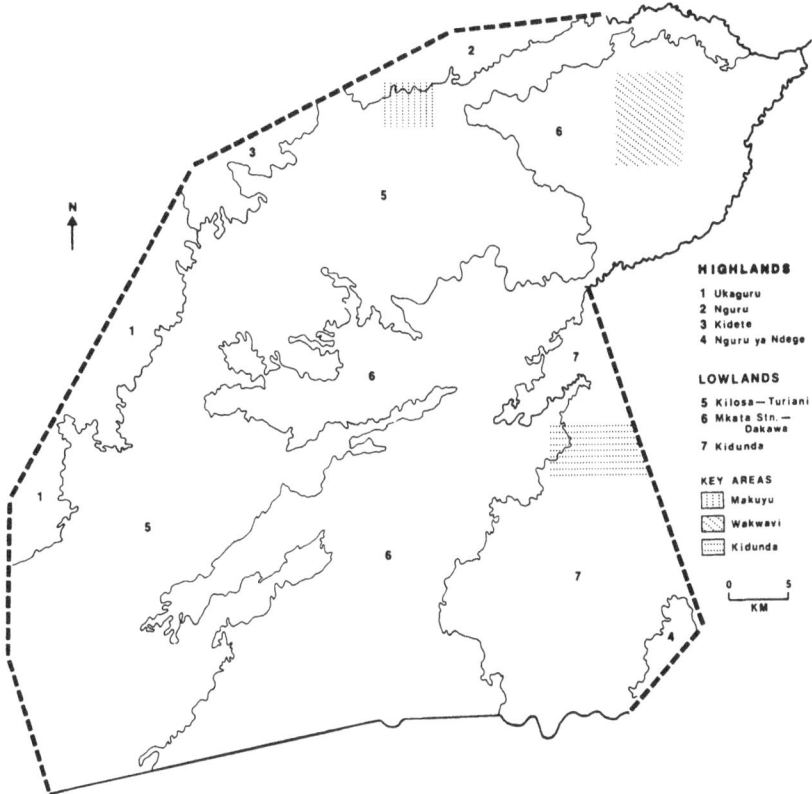

Figure 3 Land units of the North Mkata Plain.

prise four mountainous areas, the Nguru, Ukaguru (often called Kaguru), Rubeho, and Uluguru mountains. Figure 1 shows the location of these features. North of the Central Railway Line, the study area, here called the North Mkata Plain, can be subdivided into four highland and three lowland land units. These land units are listed below and mapped in Figure 3.

Highlands	Lowlands
Ukaguru	Kilosa-Turiani
Nguru	Mkata Station-Dakawa
Kidete	Kidunda
Nguru ya Ndege	

The highland land units have been distinguished principally on the basis of their location, range of elevation, and degree of dissection. Three of these units (the Ukaguru, Nguru, and Kidete) form an escarpment along the western margin of the study area. In the Wami River Basin the Ukaguru and Nguru mountains attain elevations of 2,264 metres and 2,112 metres AMSL, respectively. The study area includes only the lower slopes of these mountains, to an elevation of approximately 600 metres. The mountains are deeply dissected by many V-shaped valleys and tributary channels. The sharp ridge crests are rarely vegetated as they consist principally of bare rock. Both upper and lower valley slopes are steep. These slopes may be covered with forest, bush, or grass, but are punctuated by numerous rock outcrops.

The third member of the western escarpment, the Kidete land unit, is essentially a foothill transition zone between the Ukaguru and Nguru mountains. It has characteristics similar to those of these mountainous land units. In the Kidete area, however, ridge crests are narrow but not sharp and are frequently forest-covered; and there are fewer rock outcrops on the valley slopes which are moderately steep to steep.

The Nguru ya Ndege land unit is located in the southeast corner of the study area. It is an outlier of the Uluguru Mountains and is separated from them by the Ngerengere River Valley, which is approximately 10 kilometres wide at this point. Nguru ya Ndege marks the watershed between the Wami River Basin and the Ruvu River Basin into which the Ngerengere flows. While the Uluguru Mountains reach elevations of over 2,000 metres, Nguru ya Ndege attains its highest elevation at 1,357 metres. Its physiographic features are the same as those described for the Ukaguru and Nguru land units.

Although there are topographical differences between the three lowland land units, they are relatively minor. No sharp breaks in slope occur to distinguish the units one from another. Therefore, in addition to location and topography, vegetation characteristics have been taken into account much more in delimiting these land units than was needed in considering the highlands.

Along the eastern margin of the study area, to the west and north of Nguru ya Ndege, is the slightly undulating Kidunda land unit. In a few locations, small inselbergs rise 15 to 20 metres above the surrounding area. In this land unit, both major and minor stream channels are narrow and flat. When they do broaden into alluvial flats to the west, the stream channels become sinuous and braided. At these locations the vegetation is predominantly grassland, but most of the Kidunda land unit is dominated by a relatively dense *Combretum-Albizzia* woodland.

The dense woodland of the Kidunda land unit marks a sharp contrast between the expansive grasslands and wooded grasslands of the Mkata Station-Dakawa land unit. In this unit, the Wami and Mkata rivers and some of their tributaries meander across extremely broad, flat floodplains. In the study area, there is a downstream drop in elevation of only 30 metres over a distance of 60 kilo-metres, from 400 metres AMSL at Mkata Station to 370 metres just north of Dakawa. The Mkata Station–Dakawa land unit includes many areas that are subjected to prolonged annual flooding, and several perennial swamps occur near the Central Railway Line. Locally there are areas of gilgai microtopography.

Johnson and Tiarks (1969) described what has been called here the Kilosa-Turiani land unit as comprising the upper and lower pediment slopes of the Ukaguru-Kidete-Nguru escarpment. Both slopes are covered by alluvial and colluvial material derived from the highlands to the west. Except for the fact that the land is slightly more undulating at the base of the escarpment, there is little to distinguish between the upper and lower pediment slopes. Physio-graphically, the Kilosa-Turiani land unit is very similar to the Kidunda land unit. Unlike the latter unit, however, it has been cleared of most of its natural vegeta-tion for cultivation. As a result, the eastern margin is generally found where undulating cultivated fields abut the level grasslands and wooded grasslands of the Mkata Station-Dakawa land unit.

The character of the soils of the study area is directly influenced by the underlying bedrock only in the highland areas. Here the bedrock is close to the soil surface and numerous outcrops occur as a result of the steepness of the slopes and the erosional history of the area.

In attempting to generalize and correlate tectonisms for the Tectonic Map of Africa, the co-ordinators subdivided the structural history of the Precambrian period into four stages (Pallister, 1971). According to their map, the North Mkata Plain is bounded and underlain by rocks of the Precambrian A, the least ancient stage. In the study area, the rocks belong to the Usagaran System – a term that was substituted for 'Basement Complex' in the official *Lexique Stratis-graphique International* in 1957.

Pallister (1971) indicates that the Usagaran System has been mapped in detail only in localized areas of Tanzania and that broad areas in between remain unmapped. In the study area, the only available geological map at a scale larger than the 1:3,000,000 sheet of the *Atlas of Tanzania* (Tanzania, 1967a) is that by Fozzard (1965) at a scale of 1:125,000. This sheet covers only the southern half of the North Mkata Plain. However, the general lithology of the Usagaran System has been sufficiently described to enable us to identify the most im-portant rocks that produced the parent materials of the soils of the highland areas (see Quennall et al., 1956; Furon, 1963; Haughton, 1963; and Saggerson, 1969).

The dominant rocks of the Usagaran System are coarsely crystalline meta-morphics. The most widespread are the acid gneisses of the Ukaguru and Nguru mountains and the acid gneisses and granulites of the Uluguru Mountains.

Among the gneisses and granulites, there are fairly common intrusions of basic and ultrabasic masses as well as of mica pegmatites. Saggerson (1969) states that at least two periods of metamorphism have occurred, as many of the basic and ultrabasic rocks have been altered, for example, with the production of amphibolites. The following are among the major rock formations that have been identified by Fozzard (1965): migmatic biotite gneiss, kyanite-garnet-biotite gneiss, feldspathic micaceous magnetite quartzite, quartso-feldspathic garnet granulite, migmatic quartzo-feldspathic gneiss and granulite.

The soils of the lowlands are only indirectly related to bedrock geology. As indicated, the lowlands are composed of materials that have been deposited there, originating from the flanking highlands. A great deal of this material has come to rest here because of the reduction of river gradients resulting from the above-mentioned crustal movements.

In the Kidunda land unit, the bedrock is covered by material deposited from the Uluguru Mountains and Nguru ya Ndege. The unit is also partly made up of the material weathered in situ. By far the majority of the deposits and in situ weathering products are of a coarse sandy nature. Clays occur in localized areas only. Throughout the Kidunda land unit, the thickness of the overburden ranges from 5 to 10 metres. Similar types of material make up the Kilosa-Turiani land unit. There, however, most of the deposits are colluvial in nature and are much thicker. Although no quantitative range for the thickness of the colluvium can be stated, an open well observed by the author did not reach bedrock at 15 metres below the surface.

Thick deposits of clays little altered by pedogenetic processes occur at the surface throughout the Mkata Station-Dakawa land unit. These are underlain by layers of sand, silt, and clay in complex combinations. The depth to bedrock of these alluvial materials is extremely variable. 'A borehole at Mkata Station is reported to have reached gneiss at 182 feet (55 metres), whereas one some 3 miles (5 kilometres) to the east is reported to have entered gneiss at 59 feet (18 metres)' (Fozzard, 1965). Twenty kilometres to the west of Mkata Station 'a vertical borehole at Kimamba to 622 feet (190 metres) penetrated only unconsolidated sands and clays and revealed no signs of bedrock' (Fozzard, 1965).

CLIMATE AND WATER RESOURCES

In addition to providing the parent materials from which the soils of this area developed, the Ukaguru, Nguru, and Uluguru mountains play an important role in determining the moisture regime of the plain. During the period from October to January, the Uluguru/Nguru ya Ndege blocks intercept the southeast monsoons, thereby creating a rainshadow over the eastern margins and central portion of the plain. The nothwestern side of Nguru ya Ndege itself is also affected to some degree, as is evidenced by the semi-arid type of vegetation that it supports. In contrast, the slopes of the Ukaguru-Kidete-Nguru scarp, 'having formerly lost their soil by denudation to the plain below, now lose their water also, and that which they part with too readily by run-off and excessively free drainage appears in the subsoil at the fringe of the alluvial plain, conferring upon the cultivation and natural vegetation there a humid aspect which is not a true reflection of the overhead condition of climate' (Milne, 1947).

This pattern is not easily visualized from the climatic data presented in Table 2 for stations in or close to the study area. The Morogoro station, for example, receives rainfall in October and November in greater amounts than the lowland stations to the west and north, probably because the rain clouds that build up over the Ulugurus spill out over Morogoro, which is at the very base of these mountains. The rainshadow effect can be seen, however, when some of this information is mapped. In Figure 4b, modified from a study of the neighbouring Ruvu Basin (Jackson, 1970), isolines of mean annual rainfall are plotted. The illustration shows the concentration of rainfall over the Ulugurus and a significant decrease in the North Mkata Plain to the west and north. Although the lowest value in Figure 4b is 800 millimetres, it has been suggested that many locations in the central portion of the study area rarely receive more than 600 millimetres of rainfall annually (Dolfi, 1963).

Table 2 illustrates the seasonal nature of the precipitation of the area. A very significant dry period occurs from June to November. In this six-month period only one lowland station (Scutari Sisal Estate) records a mean monthly rainfall greater than 50 millimetres, and then in only one month (November). It is popularly held that there are two rainy periods. The first, from October to January, is called the *Short Rains* – probably because it is really only in December and January that significant amounts of precipitation are received. The second, from March to May, is called the *Long Rains*. In fact, as Table 2 demonstrates, there is really only one rainy season from December to May, the break in February being relatively minor or not noticeable at all.

In Tanzania, 800 millimetres of mean annual rainfall is frequently cited as the amount necessary to sustain permanent cultivation (Dagg, 1969). In the study area, more than this amount is received on the average only along the western margin (see Stations 3 and 4, which are the only stations of the nine plotted in Figure 4a that are actually within the study area). The central and eastern portions of the North Mkata Plain receive just this amount or less.

But mean annual figures are often misleading. They mask the seasonal distribution of the annual amounts as indicated above. Even more significantly, they do not take into consideration the dependability of the precipitation. The pre-

TABLE 2

Mean monthly and mean annual precipitation (in mm) for selected rainfall stations in or close to the North Mkata Plain

Station name:	Morogoro	Muskati Mission	Scutari Sisal Estate	Marios Sisal Estate	Mhondo Mission	Msowero	Mvomero	Masimbu Sisal Estate
Station number:	96.3700	96.3710	96.3713	96.3714	96.3718	96.3719	96.3721	96.3733
Altitude (metres):	579	1,829	457	362	487	487	487	487
Years in operation:	69	38	36	36	70	36	36	20
January	88.6	124.7	131.6	106.9	190.0	132.6	110.5	127.5
February	104.9	133.2	147.1	134.9	142.0	111.0	94.5	78.5
March	146.3	169.0	183.4	164.9	275.1	158.0	161.5	106.7
April	216.4	253.2	181.6	167.4	350.3	249.4	231.9	176.8
May	92.2	117.6	59.2	51.3	188.5	88.6	100.3	65.5
June	20.8	32.3	4.8	3.0	51.1	10.4	17.5	6.1
July	14.0	21.1	5.1	3.8	60.2	8.1	11.7	2.5
August	10.9	18.3	6.4	10.2	43.2	3.8	13.2	6.6
September	14.0	27.9	5.3	5.1	46.7	4.6	13.5	6.6
October	35.3	46.0	26.7	23.1	78.0	16.3	32.3	12.7
November	69.6	78.7	54.6	49.5	127.0	48.8	63.8	33.3
December	81.0	117.6	105.9	96.5	228.1	104.9	82.3	59.2
TOTAL	893.8	1,139.6	911.4	816.6	1,780.0	936.5	932.9	682.0

SOURCE: Compiled from data in Jackson (1970) and from P.W. Porter (private communication)

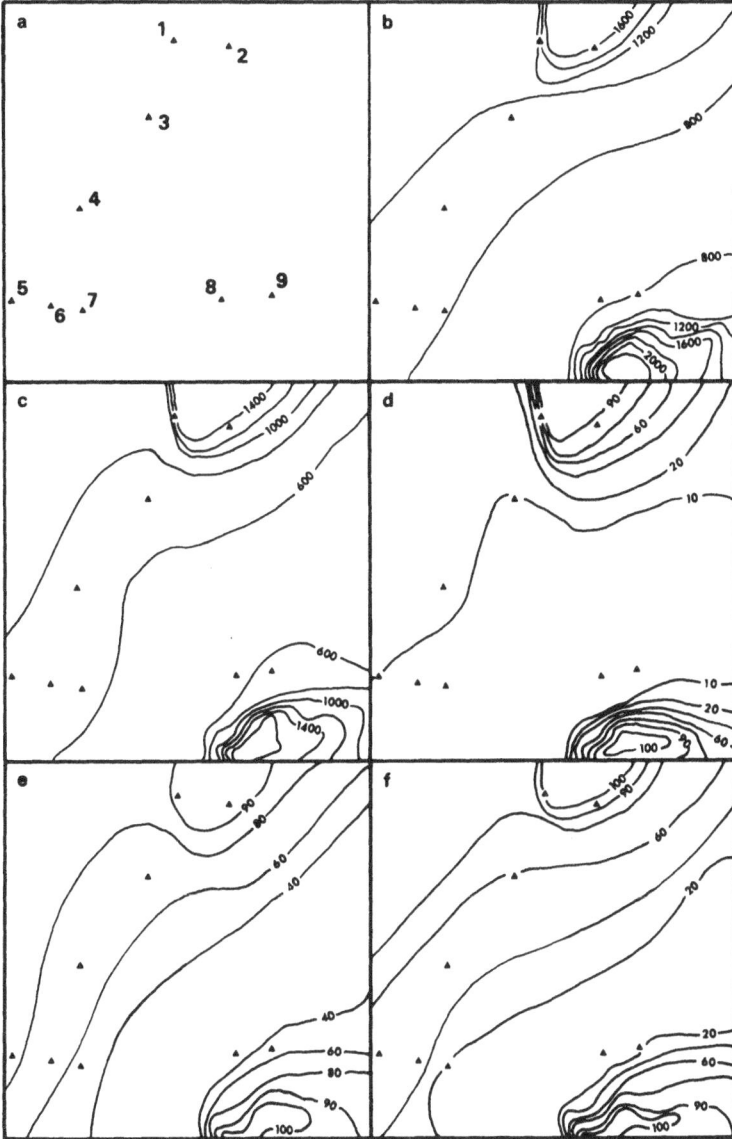

Figure 4 Rainfall distribution characteristics of the North Mkata Plain.
(a) Stations: 1, Muskati Mission; 2, Mhondo Mission; 3, Mvomero; 4, Msowero;
5, Kilosa; 6, Scutari Sisal Estate; 7, Marios Sisal Estate; 8, Masimbu Sisal Estate;
9, Morogoro. (b) Mean annual rainfall (mm). (c) Minimal annual rainfall (mm)
expected to occur with 80 per cent probability. (d) Percentage probability of
receiving at least 1,200 mm of rainfall per year. (e) Percentage probability of
receiving at least 800 mm of rainfall per year. (f) Percentage probability of
receiving at least 800 mm of rainfall in four out of five years.

carious nature of the rainfall distribution over the North Mkata Plain is demonstrated in Figures 4c to 4f. Again it is only along the western margin of the study area that there is some degree of rainfall reliability. But even there problems arise as: at the 80 per cent probability level, the receipt of only 600 to at least 800 millimetres of rainfall can be assured; the probability of receiving large amounts of annual precipitation in the order of 1,200 millimetres is 10 per cent or less; the probability of receiving the target figure of 800 millimetres of rainfall per year is only in the order of 60 to 80 per cent; and the probability of receiving 800 millimetres of rainfall in four out of five years is only 60 per cent.

Throughout Tanzania, as in the North Mkata Plain, the variable nature of the rainfall regime is considered one of the most limiting factors in agriculture. This consideration is frequently followed by discussions of irrigation possibilities. Sharma (1957) and Dolfi (1963) have suggested that much of the North Mkata Plain could be irrigated using surface water supplemented with groundwater. Irrigation development will require an examination of river discharges, water quality, water balance, and soil characteristics of the North Mkata Plain.

The river drainage pattern of this part of the Wami River Basin reflects the areal distribution of rainfall. Because of the Uluguru rainshadow (and river capture by the Ngerengere River) few streams flow from east to west into the North Mkata Plain. In the study area, only one major westbound stream channel can be traced: the Vilanza River from Nguru ya Ndege west to the Mkata River (see Figure 5). It is dry throughout much of the year and high flows are carried for only short periods during the rainy season. A number of small, unnamed stream channels do exist north of Nguru ya Ndege, but they are very narrow, shallow, and of a seasonal nature.

In the study area, the major rivers and their tributaries drain the central and western portions of the North Mkata Plain. With the exception of the Mkata River, the headwaters of these rivers can be located in the Ukaguru and Nguru mountains and in the Kidete land unit. All of these major rivers carry water throughout the year, although there are significant differences between the dry and wet season flows.

Considerable difficulties would be encountered in using these rivers for surface irrigation on a large scale. Streamflow in the eastern portion of the study area is extremely low and in no way dependable. The flows in tributaries originating in the western highlands are only slightly more reliable. Dry season discharge is low and difficult to predict; and wet season discharge is such that the Mkata and Wami rivers and their tributaries cause problems by flooding annually. For example, as a result of forest clearing and indiscriminate cultivation on steep slopes in the upper catchment areas of the Mkundi River, the area around the village of Magole is yearly subjected to serious flooding. This flooding frequently

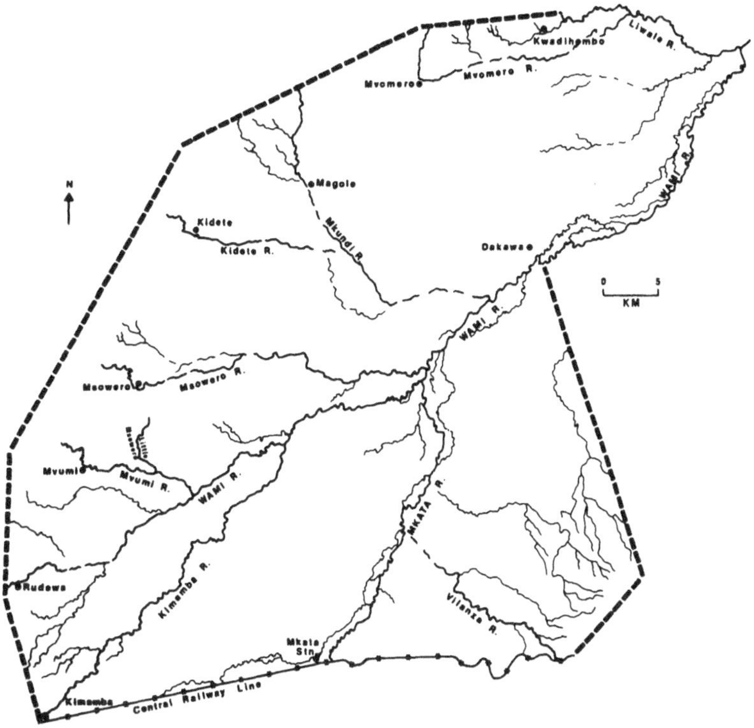

Figure 5 Main drainage patterns of the North Mkata Plain.

washes out the Kilosa-Turiani road, deposits large amounts of sand and silt on the fields, and is attendant with numerous property losses and health problems (see Illustrations 1 and 2). The use of these rivers for irrigation agriculture will require that major consideration be given to upper catchment rehabilitation. This is particularly important given that rainfall erosivity is extremely high in similar catchment environments in Tanzania. As a result, unprotected, cultivated land is subject to severe erosion (Rapp et al., 1972; Temple and Rapp, 1972).

Conductivity measurements and sodium absorption ratios are indicated in Table 3 and Figure 6 for water samples collected from rivers and boreholes in the study area. Five of these were collected by the author and analysed by staff of the Water Development and Irrigation Division (WD&ID) of the Ministry of Agriculture (Morogoro office). The remainder are from Dolfi (1963).

TABLE 3

Conductivity values and sodium absorption ratios of the waters of the North Mkata Plain

Sample number	River or borehole (borehole number in brackets)	Conductivity (micromhos/cm at 25°C)	SAR†	Water class
1*	Wami at Rudewa	60	4	C1S1
2	Wami at Mkata junction	90	5	C1S1
3*	Wami at Dakawa	175	4	C1S1
4	Mkata at Central Rail Line	110	4	C1S1
5*	Mvumi	30	4	C1S1
6*	Mkundi	390	8	C2S1
7*	Mvomero	50	2	C1S1
8	Kidete	400	12	C2S2
9	Kimamba	250	6	C1S1
10	Makuyu (open well)	760	10	C3S2
11*	Kimamba Sisal Estate (42/56)	n.a.	n.d.	C2
12*	Kimamba (31/60)	n.a.	n.d.	C2
13*	Masimbu (1/40)	n.a.	n.d.	C3
14*	Ilonga Sisal Estate (14/38)	n.a.	n.d.	C3

* Data from Dolfi (1963).
† Estimated for Dolfi's data but determined by the Morogoro office of the Water Development and Irrigation Department for the author for the present study.
n.a. Data not available.
n.d. Measurement not determined.

These measurements indicate that most of the rivers could supply water suitable for irrigation agriculture. Except for the Mkundi and Kidete rivers, the conductivity measurements of the surface waters of the North Mkata Plain are equal to or less than 250 micromhos/centimetre measured at 25°C. According to the United States Salinity Laboratory (Richards, 1954), waters having such conductivity measurements are classed as Low Salinity Water (C1) and are eminently suitable for irrigation purposes. The surface waters of the study area, except for the Kidete River, are low also in dissolved sodium (S1), likewise making them suitable for irrigation.

Over forty boreholes are in operation in the study area, principally on sisal estates and on the National Development Corporation cattle ranch near Mkata Station. Unfortunately, their salinity content (classes C2 and C3), though acceptable for human and livestock consumption, limits their use in supplying water for irrigation. Through evaporation, water from these sources would soon cause high salinity concentrations in the soils. It should be noted, as well, that

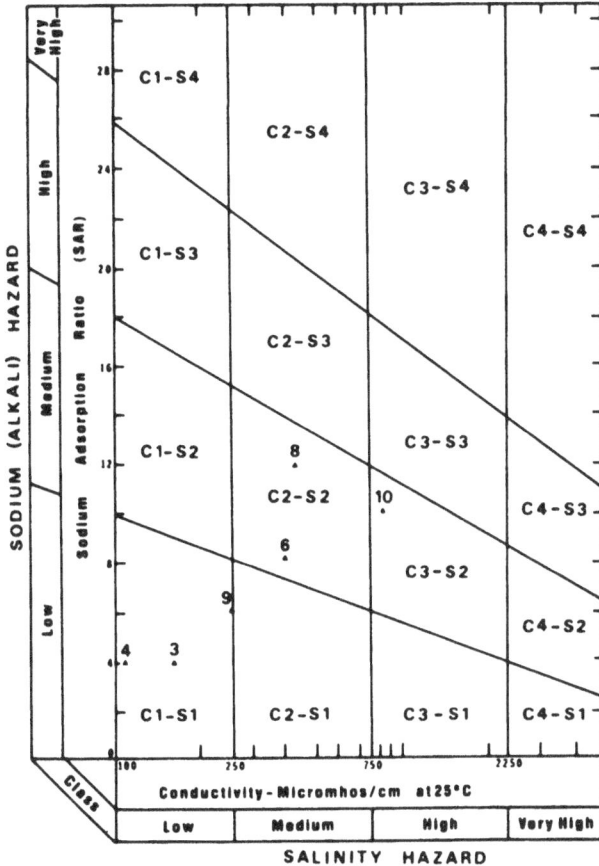

Figure 6 Irrigability of the waters of the North Mkata Plain.

the C1 surface waters could also produce high salinity concentrations if special precautions are not taken, as soils of restricted natural drainage are widespread.

In examining the water balance characteristics of the entire Mkata Plain, Dolfi (1963) assumed an area of 2,000 square miles and a yearly rainfall throughout of 30 inches. Using these data he calculated a total annual rainfall figure for the plain of 3,000,000 acre-feet. Of this, he estimated that only 200,000 acre-feet would be available for irrigation purposes as 2,800,000 acre-feet would be lost through evaporation. He felt that through proper field management and flood control 100,000 acre-feet could be conserved and when added to river inflow would give another 700,000 acre-feet of water that could be used for large-scale irrigation agriculture.

But Dolfi's conclusion that a total of 900,000 acre-feet of water could be made available for irrigation must be viewed with caution. In the early 1960s, when Dolphi made these calculations, the only figures available to him were gross estimates of stream inflow and evaporation. And furthermore he did not take into account the great temporal and spatial variations in the annual amounts of rainfall that are received through the Mkata Plain. Indeed, not much more than these gross estimates are available today. There is an exception, however. Evaporation data for several of the rainfall stations in the study area have recently become available. They allow an examination of the water balance characteristics at these locations which can be used to indicate more clearly the need for irrigation.

Water balance diagrams (Figure 7) and tables (Appendix A) have been constructed for eight meteorological stations in or close to the study area. They compare monthly values of precipitation with potential evaporation (E_o), where E_o relates the amount of evaporation from an open water surface to a number of climatological parameters such as mean air temperature, dew point temperature, run of wind, and incoming solar radiation (Penman, 1948). In constructing these diagrams and tables a soil moisture storage capacity of 250 millimetres was assumed. This figure 'corresponds to estimates for East Africa and general values based on conditions in other parts of the world' (Nieuwolt, 1973).

Nieuwolt (1973) indicates that in areas where the annual surplus of precipitation over evaporation is less than 50 millimetres 'irrigation is absolutely necessary for all forms of crop agriculture.' For all of the stations examined in this study the annual E_o values exceed the mean annual precipitation by a factor of 2 or more. In terms of monthly figures, every station has adequate rainfall to meet potential evaporation only a few months of the year, after which supplementation from soil moisture stored in previous months is required. Exhaustion of this storage leads to large annual deficits of 900 to 1300 millimetres, except at Mhondo Mission, where a small annual surplus of 100 millimetres counteracts a deficit of 300 millimetres. According to Nieuwolt's statement, to sustain crop agriculture in the North Mkata Plain would require the use of irrigation. But crops have been cultivated for decades in this area without irrigation. Two explanations may be offered to account for this fact.

Most of the arable cultivation that is carried out in the North Mkata Plain is in the Kilosa-Turiani land unit, immediately east of the Ukaguru-Kidete-Nguru escarpment. As indicated earlier in this section, this land unit has a more humid aspect than would be expected from the overhead climatic conditions. Large amounts of runoff and drainage water are received here from the steep, eroded slopes of the escarpment.

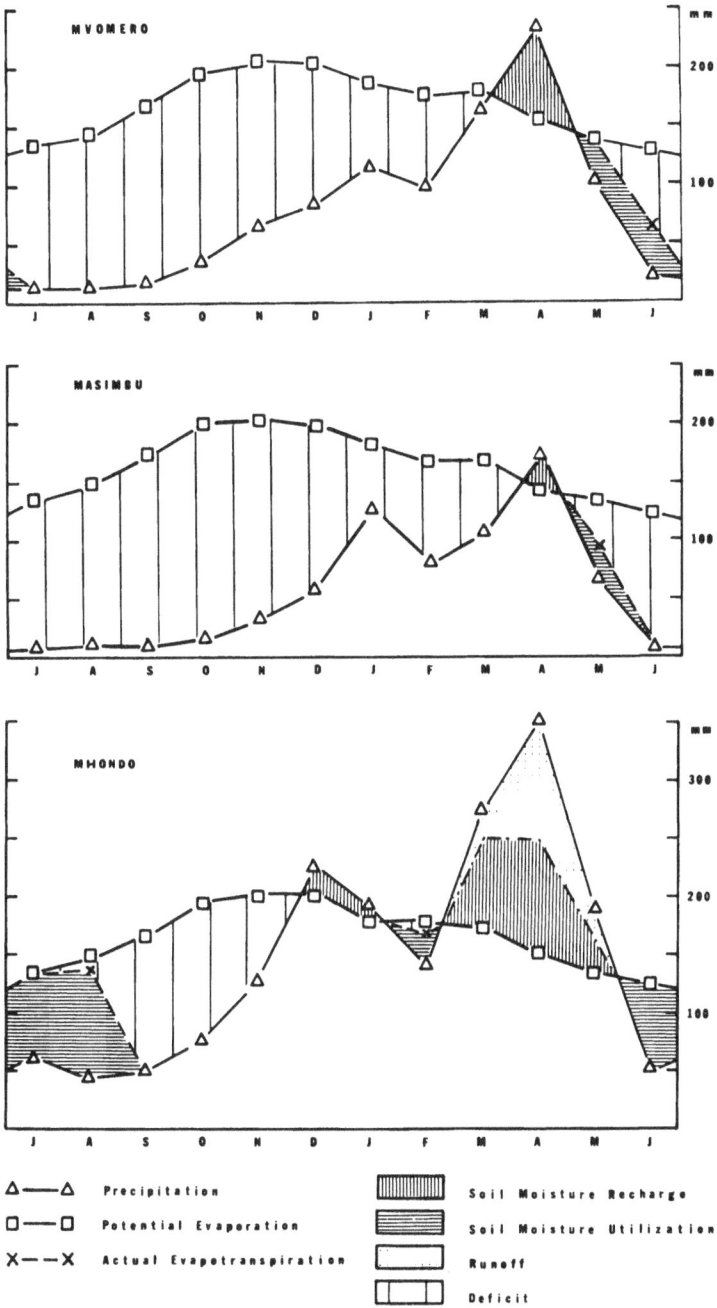

Figure 7 Water balance diagrams for Mvomero, Masimbu, and Mhondo.

A second, more fundamental explanation is the fact that potential evaporation does not measure the moisture requirements of a crop. In addition to the climatological parameters that are used in computing E_0, a crop's demand for water is dependent on the nature of the plants and the stages of their growth cycle. At times, a crop needs considerably less moisture than is indicated by the values of E_0. At other times the moisture requirements may be equal to or exceed potential evaporation (Chang, 1968). It is important then to consider the relationship between the actual water need (E_t) of a crop and the potential evaporation (E_0) as expressed by the ratio $E_t:E_0$.

It is thought that the $E_t:E_0$ ratios at different stages of growth are constant for a particular crop and do not vary with location. 'Thus, relations derived, say, in the USA could be applied with some confidence here and certainly relations determined in one part of East Africa should apply elsewhere in East Africa' (Dagg, 1965). For example, the water requirements of East African long-term maize varieties are such that the monthly $E_t:E_0$ ratio changes from 0.45 in the month the crop is planted to values of 0.48, 0.63, 0.87, 1.00, 0.92, 0.67, and 0.45 in each of the following months. For the imported varieties of maize which have a shorter growing period, the figures are 0.45, 0.49, 0.70, 0.94, 1.00, 0.75, and 0.47 (Dagg, 1965).

Because maize is a major food and cash crop of the North Mkata Plain, these figures have been used to calculate the water supply and demand requirements for long- and short-term maize varieties for two stations, Msowero and Mvomero. The values, found in Tables 4 to 7, have been calculated for four months of planting, November through February. The tables illustrate not only the difficulty in supplying adequate moisture to crops in the North Mkata Plain but also the importance of time of planting.

At Msowero, long-term varieties of maize would, according to Table 4, be supplied with adequate moisture if planted in December or January, although there would be stress at the end of the growing cycle for the crop planted in January. Crops would also be relatively well supplied with moisture if planted in February, although there is the danger of two months of water deficit at the end of the growing period. Crops planted in November would probably suffer severely as indicated by the water deficits in the month of planting and four subsequent months. A similar pattern of water supply and demand emerges in Table 5 for the short-term maize varieties with the exception that a water deficit does not occur at the end of the growing cycle when the crop is planted in January. The number of deficit months for the crop is reduced to one when it is planted in February.

Long-term maize varieties grown at Mvomero (see Table 6) will suffer from water deficits for one month both at the beginning and at the end of the growing

TABLE 4

Estimated water supply and demand for long-term maize varieties at Msowero*

		Jan	Feb	Mar	Apr	May	June	July	Aug	Sept	Oct	Nov	Dec
Precipitation (mm)		133	111	158	249	89	10	8	4	5	16	49	105
E_o (mm)		180	169	170	146	135	128	134	146	173	202	205	199
$E_t : E_o$ for four planting dates	P1	0.63	0.87	1.00	0.92	0.67	0.45					0.45	0.48
	P2	0.48	0.63	0.87	1.00	0.92	0.67	0.45					0.45
	P3	0.45	0.48	0.63	0.87	1.00	0.92	0.67	0.45				
	P4		0.45	0.48	0.63	0.87	1.00	0.92	0.67	0.45			
E_t (mm) for four planting dates	P1	113	147	170	134	90	58					92	96
	P2	86	106	148	146	124	86	60					90
	P3	81	81	107	127	135	118	90	66				
	P4		76	82	92	117	128	123	98	78			
Cumulative soil water storage (mm) for four planting dates	P1	-14	-50	-62	53	52	4					-43	-34
	P2	62	67	77	180	145	69	17					15
	P3	52	82	133	250+	204	96	14	-48				
	P4		57	86	152	250+	132	17	-77	-150			

* E_o is the potential evaporation measured from an open water body. $E_t : E_o$ is the ratio between the water needs of maize and potential evaporation. P1, P2, P3, and P4 are the planting dates November, December, January, and February, respectively. When there is a negative sign before a figure it indicates that there will be a water deficit; a positive sign after a figure indicates that there will be runoff.

TABLE 5

Estimated water supply and demand for short-term maize varieties at Msowero*

		Jan	Feb	Mar	Apr	May	June	July	Aug	Sept	Oct	Nov	Dec
Precipitation (mm)		133	111	158	249	89	10	8	4	5	16	49	105
E_o (mm)		180	169	170	146	135	128	134	146	173	202	205	199
$E_t : E_o$ for four planting dates	P1	0.70	0.94	1.00	0.75	0.47						0.45	0.49
	P2	0.49	0.70	0.94	1.00	0.75	0.47						0.45
	P3	0.45	0.49	0.70	0.94	1.00	0.75	0.47					
	P4		0.45	0.49	0.70	0.94	1.00	0.75	0.47				
E_t (mm) for four planting dates	P1	126	159	170	110	63						92	98
	P2	88	118	160	146	101	60						90
	P3	81	83	119	137	135	96	63					
	P4		76	83	102	127	128	101	69				
Cumulative soil water storage (mm) for four planting dates	P1	-57	-105	-117	22	48	106					-43	-50
	P2	60	67	65	168	156	99	44					15
	P3	52	80	119	231	185	132	39					
	P4		56	85	142	250+	132	39	-29				

* E_o is the potential evaporation measured from an open water body. $E_t : E_o$ is the ratio between the water needs of maize and potential evaporation. P1, P2, P3, and P4 are the planting dates November, December, January, and February, respectively. When there is a negative sign before a figure it indicates that there will be a water deficit; a positive sign after a figure indicates that there will be runoff.

TABLE 6

Estimated water supply and demand for long-term maize varieties at Mvomero*

		Jan	Feb	Mar	Apr	May	June	July	Aug	Sept	Oct	Nov	Dec
Precipitation (mm)		111	95	162	232	100	18	12	13	14	32	64	82
E_o (mm)		186	176	174	151	136	127	134	143	169	194	203	202
$E_t : E_o$ for four planting dates	P1	0.63	0.87	1.00	0.92	0.67	0.45					0.45	0.48
	P2	0.48	0.63	0.87	1.00	0.92	0.67	0.45					0.45
	P3	0.45	0.48	0.63	0.87	1.00	0.92	0.67	0.45				
	P4		0.45	0.48	0.63	0.87	1.00	0.92	0.67	0.45			
E_t (mm) for four planting dates	P1	117	153	174	139	91	58					91	97
	P2	89	111	151	151	125	86	60					91
	P3	84	84	110	131	136	118	90	64				
	P4		76	82	92	117	128	123	98	78			
Cumulative soil water storage (mm) for four planting dates	P1	-48	-106	-118	-25	-16	-56	-20				-27	-42
	P2	13	29	40	121	96	28	-20					-9
	P3	27	38	90	191	155	55	-23	-74				
	P4		16	94	231	223	114	3	-80	-142			

* E_o is the potential evaporation measured from an open water body. $E_t : E_o$ is the ratio between the water needs of maize and potential evaporation. P1, P2, P3, and P4 are the planting dates November, December, January, and February, respectively. When there is a negative sign before a figure it indicates that there will be a water deficit; a positive sign after a figure indicates that there will be runoff.

TABLE 7

Estimated water supply and demand for short-term maize varieties at Mvomero*

		Jan	Feb	Mar	Apr	May	June	July	Aug	Sept	Oct	Nov	Dec
Precipitation (mm)		111	95	162	232	100	18	12	13	14	32	64	82
E_O (mm)		186	176	174	151	136	127	134	143	169	194	203	202
$E_t : E_O$ for four planting dates	P1	0.70	0.94	1.00	0.75	0.47						0.45	0.49
	P2	0.49	0.70	0.94	1.00	0.75	0.47						0.45
	P3	0.45	0.49	0.70	0.94	1.00	0.75	0.47					
	P4		0.45	0.49	0.70	0.94	1.00	0.75	0.47				
E_t (mm) for four planting dates	P1	130	165	174	113	64						91	97
	P2	91	123	164	151	102	60						91
	P3	84	86	122	142	136	96	63					
	P4		79	85	106	128	127	101	67				
Cumulative soil water storage (mm) for four planting dates	P1	-61	-131	-139	-20	16	18					-27	-42
	P2	11	-17	-19	62	60	18						-9
	P3	27	36	76	166	130	52	1					
	P4		16	93	219	203	94	5	-49				

* E_O is the potential evaporation measured from an open water body. $E_t : E_O$ is the ratio between the water needs of maize and potential evaporation. P1, P2, P3, and P4 are the planting dates November, December, January, and February, respectively. When there is a negative sign before a figure it indicates that there will be a water deficit; a positive sign after a figure indicates that there will be runoff.

period when planted in December. Even when the crops are planted in January or February there will be a two-month period of moisture deficiency. Table 7 indicates that the best month for planting short-term varieties of maize is January. No deficits of moisture occur in any of the months of the growing period. February would also be a fairly good planting month as the crops would experience a deficit of moisture only when they are near harvest stage. But short-term and long-term varieties of maize would suffer severe moisture deficits if planted in November.

The analyses of Figure 7 and Tables 4 to 7 and of the additional data provided in Appendix A suggest that crops adapted to drought conditions, or which have water-demanding patterns like those of maize, can be grown in many parts of the North Mkata Plain having similar climatic characteristics. They also show that the margin for error in planting dates is small. This margin could be increased with the provision for irrigation.

There is, however, one major drawback in this analysis, although it supports even more the case for irrigation. Throughout Tanzania, the values for annual and monthly rainfall are positively skewed (Nieuwolt, 1973). That is, the median values are lower than the mean values. As the monthly rainfall data used in Appendix A are mean values, they overestimate the amount of moisture that would be available to crops in most years. Supplementary irrigation would therefore be very beneficial.

Without irrigation, planting dates become even more critical than the discussion above indicates. December planting, for example, would be risky at Msowero as the 15 millimetres of moisture which shows as a surplus in Tables 4 and 5 may not materialize in most years. The time-of-planting data reviewed by Akehurst and Sreedharan (1965) confirm these general observations. In the experiments conducted at Ilonga (just outside the survey area), maize yields were greatest when planting was done in January or February, with variations dependent on the pattern of rainfall received in the experimental years. Similar conclusions could be drawn from their yield data for groundnuts, soyabeans, cotton, and castor (Akehurst and Sreedharan, 1965).

It is obvious from this discussion of the water balances that irrigation would be a beneficial, if not necessary, aid to agriculture in the North Mkata Plain. In fact, Dolfi (1963) concluded that sufficient water was available for this purpose. He also concluded that the development of large-scale irrigation projects awaited only the provision of flood control works and the construction of a network of roads to allow access to many of the areas that are now difficult to reach. Unfortunately, the task will not be as simple as this. One major element missing from Dolfi's study was the analysis of the soils. As will be shown in later chapters, more than half of the North Mkata Plain has soils that may be irrigated only with great difficulty and expense.

VEGETATION

The native vegetation of the study area has been greatly modified by man's activities. These have included cultivation, grazing, and burning. In general, the vegetation now consists of fire-maintained grasslands and wooded grasslands in the lowlands and natural or semi-natural woodlands and forests in the highlands.

On the lower, drier slopes of the Ukaguru, Kidete, Nguru, and Nguru ya Ndege land units, *Combretum-Brachystegia* woodland (*miombo*) dominates (see Illustration 3). The most important tree species are *Combretum molle, Brachystegia longifolia, B. bachmii,* and *Julbernardia globiflora.* Also represented are *Acacia tortilis* and *Euphorbia candelabrum.* On the upper slopes of the Ukaguru and Nguru mountains where rainfall is heavy, a moist forest of mixed coniferous and deciduous species predominates. Little of this forest occurs in the study area.

Up to 90 per cent of the Kilosa-Turiani unit is cultivated, with sisal estates in the central and southern sections and large mechanized maize-sunflower-millet farms to the north. Sugarcane farms are also found in the northern as well as the central sections. Scattered throughout are a large number of subsistence and semi-subsistence farms growing maize, millet, cotton, castor, cassava, bananas, and other crops. In the uncultivated areas the main tree species are *Sclerocarya* spp. and *Acacia polycantha.* The most common grasses are *Panicum maxima, Hyparrhenia* spp., and *Pennisetum purpurem.*

Only a small portion of the Mkata Station-Dakawa unit is cultivated. This includes the isolated subsistence farms along the Central Railway Line and some north of the Wakwavi Settlement near Dakawa. The semi-nomadic Wakwavis graze cattle in many parts of this land unit, especially in the area north from Msowero to the Wakwavi Settlement. In the south is a large government cattle ranch using the plain as unimproved pasture. Where there is no cultivation, the vegetation is open wooded grassland and grassland that is maintained by fire (see Illustration 4). The most common trees in this land unit are *Acacia nigrescens, Combretum apiculatum, Terminalia spinosa,* and *Spirostachys africana,* The main grasses are *Themeda traindra, Hyparrhenia* spp., and *Digitaria* spp.

Little of the Kidunda unit is used for cultivation. There are a few scattered subsistence plots and only one sisal estate. Although there are some significant grassland areas, Combretum-Albizia woodland is dominant. The most important woodland species are *Combretum molle, Acacia nigrescens, Albizia* spp., and *Sclerocarya* spp. *Panicum* spp. and *Hyparrhenia* spp. are the common grasses.

KEY AREAS

The three key areas that have been referred to in the text are roughly outlined in Figure 3. They have been labelled Makuyu, Kidunda, and Wakwavi.

The Makuyu key area is located principally in the Kilosa-Turiani land unit, but overlaps slightly the boundary between this unit and the Nguru land unit. Sixteen square kilometres in area, the Makuyu key area is almost entirely cleared for cultivation. Both small- and large-scale agricultural holdings may be found here, but the former dominate. The area is accessible from both the Kilosa-Turiani and Morogoro-Mvomero roads. Access to the northern and western portions of the area can also be gained via the newly widened Mvomero-Kibati road, which runs along the base of the Nguru Mountains at this location.

At the time of the field survey, only three smallholdings were found in the Kidunda key area. This is a heavily wooded area covering approximately twenty-eight square kilometres of the Kidunda land unit. Grassland and wooded grassland sections may be found locally distributed throughout the area as well as on its western margin, which overlaps the Mkata Station-Dakawa land unit. Access to the Kidunda key area may be gained from the Morogoro-Mvomero road by two dirt tracks, impassable during the wet season.

The Wakwavi key area, covering close to thirty-eight square kilometres, is situated entirely within the Mkata Station-Dakawa land unit. It is accessible via a dirt track leading from the Dakawa-Kwadihombo road. This is the site of the Wakwavi Settlement Scheme, under construction at the time of the survey. Except when under flood waters, the grasslands and wooded grasslands of the area are periodically grazed by cattle belonging to the Wakwavi people.

PART II

Soils and Land Capability

4

Soils of the North Mkata Plain

The soils of the North Mkata Plain are described in the following pages by first grouping the soil series under headings similar to those of the *Soil Map of Africa* (D'Hoore, 1964). Then the series are described in terms of their morphological and analytical profile characteristics. These descriptions summarize the more detailed data that are found in the appendices. Inferences with respect to soil genesis are made by reference to these profile data and by correlation of the soil units with some of the classification categories of *Soil Map of Africa* (D'Hoore, 1964) and *Soil Taxonomy* (Soil Survey Staff, 1975). The tables in Appendix A can be used for reviewing some of the more significant analytical and morphological differences between the various soil series.

VERTISOLS OF TOPOGRAPHIC DEPRESSIONS

In Tanzania, *vertisolic soils* are frequently found in topographic depressions into which base-rich waters are draining. These dark clay soils are characterized by the dominant presence of 2:1 expanding lattice clays, a very high cation exchange capacity (CEC), and a high percentage base saturation. As the 2:1 clays are primarily montmorillonitic, these soils commonly expand and contract greatly with alternate wetting and drying. This process gives rise to features such as self-mulching, deep cracking in the dry season, and a microtopography of small hummocks called *gilgai*. Throughout East Africa, these soils are popularly known as *black cotton soils* or *mbuga soils*. Both are misnomers, as these soils are not necessarily suitable for cotton, and *mbuga* is a term that more properly describes seasonally flooded valley grasslands where vertisols may or may not be present. Vertisols are widespread in the North Mkata Plain and are here represented by three soils series: Mkata, Kwavi, and Kwadihombo.

Mkata Series

The Mkata series is composed of deep, fine-textured soils that have developed on lacustrine and alluvial sandy clays. Natural drainage is impeded because of the high clay content (exceeding 40 per cent throughout the profile) and poor structure, which changes from coarse subangular blocky to structureless and massive with depth. These soils occur on land having slopes of 1 per cent or less and are flooded for moderate to prolonged periods annually. Although there has been very little erosion, the features described above cause these soils to be placed in class 4 of the capability scheme as defined in Chapter 2.

Although no mineralogical analysis has been conducted for these soils, their adjusted CEC:clay ratios are greater than 0.50 and increasing with depth, which indicates that they are dominated by 2:1 lattice clays. Despite the fact that the parent materials of these soils are derived from rocks that are mainly acidic, 2:1 lattice clay formation is not surprising here. In the process of weathering, the rocks of the Nguru, Kaguru, and Uluguru mountains have released appropriate kinds and amounts of materials – notably silica, magnesium, and calcium – for such genesis to occur. As indicated below, magnesium is in plentiful supply, as is calcium, which enables the soils to maintain a high pH. It has already been indicated that the soils are dense and drainage is restricted as a result. All of these features fulfil conditions necessary for this type of clay formation (Buringh, 1970).

Because of the high 2:1 lattice clay content and the alternating wet and dry seasons, the soils of the Mkata series exhibit many features typical of vertisols – a high CEC, cracking and swelling, self-mulching, and gilgai relief. The CEC is in fact very high, ranging from over 30 milliequivalents per 100 grams of soil to over 90 me/100 g in the subsoil. The exchange complex is dominated by bases to the extent that excess bases, particularly calcium, precipitate out in the form of concretions which may be up to 5 centimetres in diameter.

In addition to accumulating calcium and magnesium, the soils of the Mkata series contain significant amounts of sodium. They tend to become alkaline with depth, and below approximately 100 centimetres the sodium percentage of the CEC greatly exceeds the 15 per cent criterion used to distinguish between an alkali and non-alkali soil, amounting, for example, to 30 per cent (Richards, 1954). Any attempt to irrigate these soils would face serious difficulties on this account.

Organic matter contributes little to the high CEC. Only in the surface horizon does the organic matter content (organic C x 1.72) exceed 1 per cent, a very low figure. The low C:N ratios of 10 and less indicate that the organic matter of these soils is rapidly mineralized. The organic matter may, however, be the cause of the dark colours throughout the profile. Although the amount is low, the organic matter is considered to be thoroughly mixed with the clays and in a calcium-rich environment is felt to impart dark colours to vertisols (Buringh, 1970). The presence of iron sulphide and manganese (although the content is low in the soils under discussion) is also felt to be partly responsible for the black colours of these organic matter-deficient soils (Dias et al., 1959).

Cracking and swelling, self-mulching, and gilgai relief are related features resulting from the process of seasonal expansion and contraction of the 2:1 clays, with alternate wetting and drying. The results of this process may be observed in the soils of the Mkata series which crack deeply in the dry season (with cracks up to 5 centimetres wide and over 100 centimetres deep) and have a thin, granular surface mulch (see Illustrations 5 and 6). Frequently some of the mulched surface material is washed or blown into these cracks. When the soils are once again moistened in the wet season, they swell and the resulting pressures produce gilgai. Gilgai relief occurs only in localized areas of the Mkata series. Where it does occur, the hummocks are less than 50 centimetres high and the distance between hummocks is often less than 1 or 2 metres.

From the data presented in Appendix C, it would appear that plants growing in the soils of the Mkata series are more than adequately supplied with plant nutrients such as potassium, magnesium, and calcium. However, a number of potential mineral disequilibrium conditions are present in these soils. Each of these is related to excessive magnesium in relation to calcium and potassium. In the surface soils, for example, the ratios Mg:K, Ca:Mg, and Ca + Mg: K are approximately 11, 3, and 46 respectively. In general, these ratios are considered satisfactory for most crops when they are 3, 5, and below 25, respectively (Boyer, 1972). In the case of the Mkata series, the soils provide a plant environment that has excessive amounts of magnesium and a deficiency of potassium. It should be noted that potassium deficiency is common to soils containing large amounts of 2:1 lattice clays. This type of clay, especially montmorillonite, is known to fix potassium.

With respect to the two other major plant nutrients, nitrogen and phosphorus, both are in short supply in these soils. It would seem that crops grown in the soils of the Mkata series should respond readily to NPK fertilizers. Field trials are necessary to confirm this though, as there is always the danger of further potassium fixation. In addition, nitrogen may be made temporarily unavailable to plants if supplied to the soil in the form of ammonia. As ammonia has a similar ionic size to potassium the 2:1 lattice clays of the Mkata soils may fix it by mechanisms similar to those that retain potassium (Page and Baver, 1940).

The features that are characteristic of the soils of the Mkata series and that which have been described above are common to the soils of the vertisol order of the Soil Taxonomy system. The Mkata series can be further classified as *typic pelloxerert*. These are vertisols that have cracks that open and close once each year and remain open for 60 consecutive days or more, occur in regions with pronounced wet and dry periods, have dry and moist chromas of less than 1.5 to a depth of 100 centimetres, and have colour values of less than 3.5 when moist and less than 5.5 when dry throughout the upper 30 centimetres of the profile.

Kwavi Series

The soils that make up the Kwavi series are deep and fine-textured, and have developed on fine lacustrine and alluvial clays. Poor structure and consistence and a clay content that approaches or exceeds 50 per cent cause these soils to be poorly drained. Areas where these soils are found have slopes of 2 per cent or less and are inundated by flood waters for moderate periods annually. In localized areas, surface erosion occurs where there has been overgrazing, and small erosion channels occur along the paths used by cattle (see Illustrations 7 and 8). As erosion of this type is not extensive, and because of drainage and inundation features, the Kwavi soils have been placed in capability class 3.

The soils of the Kwavi series, like the Mkata soils, exhibit features common to vertisols. They are self-mulching and crack during the dry season (with cracks 2 to 3 centimetres wide and in some places 60 centimetres deep), and gilgai relief is evident but not widespread. In these black clays with a thin (10 centimetre) surface horizon of sandy clay, 2:1 expanding lattice clays dominate. Their genesis parallels that of the Mkata series soils. Calcium and magnesium dominate the base-saturated exchange complex; CECs are high and increase with depth. Most soils of this series have small amounts of carbonate concretion accumulations beginning at 60 centimetres below the surface.

In general these soils have absolute amounts of magnesium that are greater, and of calcium that are less, than in the soils of the Mkata series. More significantly, there is both an absolute and a relative increase in potassium content. Little mineral disequilibrium is likely to be present that could not be overcome fairly easily with K fertilizer as the Ca + Mg: K ratios rarely exceed 25, and when they do the excess is minimal. Organic matter is rapidly mineralized as evidenced by the very low values for organic matter content and C:N ratios. Crops should respond to NP fertilizers, as nitrogen percentages are very low, and phosphorus is in medium supply only in the surface soils.

The poor drainage characteristics of these soils and the presence of significant amounts (20 per cent of the CEC) of exchangeable sodium at depths below 60 centimetres would seriously restrict the use of these soils for irrigation.

According to the Soil Taxonomy system, these soils would be classified as typic pelloxererts.

Kwadihombo Series

The vertisolic soils of the Kwadihombo series are lighter in colour than those of the previous two series. Although the colours are still dark (10YR and 7.5YR), values are greater (3 and 4). The soils are deep, fine-textured, imperfectly drained, and have formed on old alluvium that is not now receiving fresh material annually. Although inundation does occur, it occurs infrequently and for short durations only. The soils are not eroded and occur on land having slopes that are generally less than 1 per cent. The Kwadihombo series has been classified as class 3 of the capability scheme.

These very dark gray to dark brown soils are dominated by 2:1 lattice clays. Throughout their profiles, adjusted CEC:clay ratios exceed 0.50 and the CEC at all depths is greater than 40 me/100 g. Although the Kwadihombo soils are base-saturated, particularly by calcium and magnesium, no concretions were evident in any of the soils examined. Their absence is somewhat unusual as there is evidence of a fluctuating water table with mottles, some of which are reddish to yellowish-red in colour, occurring between the depths of 10 and 80 centimetres below the surface. Anderson (1963) suggests that the non-calcareous vertisols of this type occur when incoming and outgoing calcium just balance.

Although it is uncertain whether this is indeed the process that is going on in the soils of the Kwadihombo series, a possible indication of it may be the uniform presence throughout the profile of between 22 and 24 me/100 g of exchangeable calcium. However, the CEC increases significantly with depth and the base saturation decreases slightly. It is highly probable that the calcium that cannot be absorbed by colloidal material in the surface horizons remains in solution until it can be absorbed by the colloidal material in the subsoil. For example, the analytical data provided for these soils in Appendix C indicate that the soil between 80 and 140 centimetres would be capable (by rough estimate) of receiving 57.47 x 0.04 plus 70.66 x 0.02 or 3.71 me/100 g of calcium from the surface soil.

The plant environment that the soils of the Kwadihombo series provides is well supplied with absolute amounts of potassium, magnesium, and calcium. However, crop growth may be adversely affected by the relative amounts of magnesium and potassium. Throughout the profiles of these soils the Mg:K ratio greatly exceeds the value above which magnesium may cause problems (i.e., greater than 3). Only in the upper 10 centimetres can the relative amount of potassium be considered suitable. Potassium deficiency is quite marked below 10 centimetres, where the Ca + Mg: K ratio is double that considered suitable for most crops. The Kwadihombo soils also contain little nitrogen or phosphorus.

Another factor that will limit the development of these soils for cultivation is their tendency to accumulate sodium. In many cases the sodium percentage of the CEC is 20 as close to the surface as 30 centimetres.

Despite all the features described above that are common to vertisols, physical characteristics such as those caused by alternate wetting and drying of 2:1 lattice clays are not outstanding. In the course of the field survey, no gilgai microtopography was observed in areas where these soils occur; and cracking and self-mulching features were relatively insignificant, although present.

It has been stated earlier that it is not so much the amount of organic matter that imparts the dark colours to vertisols as the thorough mixing of the organic matter with clay in a calcium-rich environment. In many ways, a comparison between the Kwadihombo soils and those, for example, of the Mkata series supports this statement. Both series provide a calcium-rich clay environment. But the Kwadihombo soils which mix little by cracking and self-mulching are lighter than the well-mixed Mkata soils despite the fact that they contain more organic matter. They are lighter by as much as 2 colour values and 1 chroma, differences significant enough to be used as differentiating criteria in the Soil Taxonomy system. Using the nomenclature of the classification system, the soils of the Kwadihombo series are *aquic chromoxererts*. They are vertisols having cracks that open and close once each year and remain open for 60 consecutive days or more; occurring in regions with pronounced wet and dry periods; having moist chromas of 1.5 or more in the upper 30 centimetres; having colour values in the upper 30 centimetres that are less than 3.5; and having prominent mottles within 50 centimetres of the surface.

EUTROPHIC BROWN SOILS

The eutrophic brown soils of the North Mkata Plain have formed on the alluvial and colluvial sand and clay deposits that are found at the base of the Ukaguru-Kidete-Nguru escarpment. In this location, with high temperatures and mean annual precipitation values greater than 900 millimetres, conditions are favourable to rapid soil formation. However, these *eutrophic brown soils* are considered to be young soils as they are rejuvenated by fresh materials eroded from the slopes of the Ukaguru-Kidete-Nguru escarpment. The resulting soils have appreciable amounts of 2:1 expanding lattice clays and a high CEC, and are more than 50 per cent base-saturated. If unrejuvenated by fresh soil material, these soils would likely lose their high base status through leaching and have a lower CEC with the transformation of 2:1 clays to 1:1 type clays (Anderson, 1963).

In the North Mkata Plain, these soils are represented by three series: Makuyu, Ilonga, and Kipinde. They are the *chocolate loams* described by Lock (1969) and the *eutrophic brown soils on alluvial deposits* described by Anderson (1967a) for this area.

Makuyu Series

The soils of the Makuyu series are found on slopes of 3 per cent on sandy clay colluvial material derived particularly from the Nguru Mountains. These dark brown to dark reddish brown soils are medium in texture (sandy clay loam throughout the profile) and well drained. Inundation occurs for very short periods only where the soils occur in proximity to stream channels whose banks may be breached in the wet season. Moderate sheet erosion occurs on cleared cultivated fields. The soils are given a capability rating of class 2.

The soils of the Makuyu series have pH values that are close to neutral (6.7 throughout the profile) and high cation exchange capacities especially to a depth of approximately 50 centimetres. The high values of the CEC are not the result of organic matter content, which is less than 2 per cent, but may be accounted for by the presence of significant amounts of 2:1 lattice clays. These clays accumulate bases, particularly magnesium and calcium. The former element is present in such quantities that it makes up more than 40 per cent of the CEC throughout the profiles. In no instance does sodium accumulate to a degree that it makes up more than 1 per cent of the CEC.

These soils are well supplied with phosphorus and potassium. Both nutrients are in plentiful supply in the upper 15 centimetres but may decrease rapidly with depth. Heavy cropping would likely deplete this presently adequate supply, however, with a resulting need for PK fertilizer. The supply of nitrogen is very low. In the surface of these soils, for example, a C:N ratio of 12.3 is obtained with values for organic carbon and nitrogen being 0.49 per cent and 0.04 per cent, respectively. All of these values are low and the C:N ratios decrease rapidly with depth, indicating a very high rate of mineralization. Overall, crops grown on the Makuyu soils would likely respond readily to applications of NPK fertilizer.

In the process of weathering of these soils and of the soils on slopes immediately above them, iron is released into solution. In some cases there is a deposition of this iron in the lower part of the profiles of the Makuyu soils. The deposited iron takes the form of small pisolitic concretions which are less than 5 millimetres in diameter. These round, pealike concretions may occupy up to 2 per cent of the volume of the subsoil. In some instances they may be found in smaller volumes as close to the surface as 15 or 20 centimetres.

D'Hoore (1968) indicates that these relatively immature, brownish soils fall mainly into the *inceptisol* order of the Soil Taxonomy system. The inceptisolic Makuyu soils can be further categorized as *typic ustropepts*. These are inceptisols that occur in warm regions, have a base saturation of 50 per cent or more, are sufficiently deep that no lithic contact is made within 50 centimetres of the surface, have a CEC of at least 24 me/100 g of clay, have up to 35 per cent clay with montmorillonitic mineralogy, and have horizons that are dry for 90 cumulative days or more in most years.

Ilonga Series

The Ilonga series is composed of soils having very dark gray to dark brown colours in the upper 50 centimetres of their profiles and underlain by dark yellowish brown materials. The parent materials consist of sandy and clay colluvium derived for the most part from the Kaguru Mountains. The medium texture of these soils is uniform throughout their profiles and they may be classed as sandy clay loams. Despite the clay content and subangular blocky structure which develops at depths of between 20 and 50 centimetres, these soils are well drained. They occur on slightly undulating land with slopes up to 4 per cent. As they are rarely flooded and suffer from only slight sheet erosion, they have been placed in capability class 1.

These are relatively fertile soils. The moderately high cation exchange capacities of more than 20 me/100 g in the horizons above 40 centimetres are dominated by bases, particularly calcium. With Ca + Mg: K ratios of less than 25, the Ilonga soils are not deficient in potassium. Mg:K ratios of from 3.70 and 6.23 in the surface soil indicate that excessive magnesium may be a slight hazard. The soils are well supplied with phosphorus, and nitrogen is present in adequate amounts.

Despite the relative fertility of the Illonga soils, yields of maize on these and other soils in the area have for a long time been rather disappointing. In many instances, the poor yields have been due to poor crop husbandry practices – particularly improper date of planting and improper plant spacing (Bolton 1971a; 1971b). A more fundamental problem, however, derives from the fact that the varieties of maize that have been grown are not particularly adapted to the environment of the North Mkata Plain.

As maize is one of the major food and cash crops of the region, a good deal of effort has been put into breeding superior maize varieties for the area. This research, conducted primarily at the agricultural experimental centres at Ilonga and Ukiriguru, has resulted in the breeding of a number of composite maize varieties. Made up of a proportion of Central American material as well as local East African maizes, the Ilonga composites in field trials can outyield farmers' maize by a factor of three (3,000 kilograms per hectare compared to 1,000 kilograms per hectare) (Bolton, 1971a). This increased productivity can be achieved without fertilizers if the plants are spaced properly and planted at the correct time. Unfortunately the composite Ilonga maize does not respond to fertilizers. The lack of response may be due to nutrient fixation by colloidal clays, the leaching of nutrients before the crop is able to use them, or improper fertilizer application. However, Kesseba and Uriyo (1971) suggest that 'the consistent lack of any significant responses with the Ilonga composite maize to NP and K application at so many sites in the Morogoro Region must raise some serious doubts on the genetic potential of this variety to show any response to fertilizer application.'

In the nomenclature of the Soil Taxonomy system, the soils of the Ilonga series are classified as *typic ustropepts*.

Kipinde Series
The sandy clay loams that make up the Kipinde series are dark brown in the surface 60 centimetres, but become dark yellowish brown at lower depths. Areas where these soils are found may be undulating, and have slopes of up to 3 per cent. Little erosion of these soils has occurred although fresh alluvial, or colluvial, sandy materials may be found deposited on their surfaces. Such deposits are especially evident near stream channels. Some subangular blocky structure occurs at depths of below 30 centimetres, and below 60 centimetres the soils are mottled and have small accumulations of pisolitic iron. These imperfectly drained soils have been placed in capability class 2.

These calcium- and magnesium-rich soils, with pH values between 7.7 and 8.0, are dominated by 2:1 lattice clays. As a result, the soils of the Kipinde series have moderately high cation exchange capacities that range from 20 to 30 me/100 g. They are base-saturated with values that are near or exceed 90 per cent. While absolute amounts of potassium appear to be sufficient for most crops, especially in the upper 30 centimetres of the soils, the Mg:K and Ca+Mg: K ratios indicate that the relative amounts of potassium are low.

Phosphorus is in plentiful supply and it would appear that there are adequate amounts of nitrogen. However, the low amounts (less than 4 per cent) of organic matter that occur in the Kipinde soils are quickly mineralized as indicated by the low C:N ratios (approximately 14 in the upper 30 centimetres and 8 or less below this depth). Much of the nitrogen that is released from the organic matter would be leached out of the soils during the wet season. Crops grown in the soils of the Kipinde series would likely respond to NK fertilizer if applied in adequate amounts and at the proper time of year.

Although the Kipinde soils have most of the characteristics of typic ustropepts, they have a slightly higher organic matter content and mottles appear within 150 centimetres of the surface. Consequently they can be classified as *aquic fluventic ustropepts*.

FERRALLITIC SOILS, FERRISOLS, AND FERRUGINOUS TROPICAL SOILS

D'Hoore (1964) estimates that soils under these headings cover approximately 32 per cent or 9,504,000 square kilometres of the total surface of the African continent (respectively, 18 per cent, 3 per cent, and 11 per cent). In Tanzania, they probably cover more than 50 per cent of the total surface. These soils are subject to the process of ferrallitization. In an acidic environment that is moist at least during some part of the year, silica and bases are rapidly leached, leaving clay-rich soils in which oxides of iron and aluminum have accumulated. Although not a diagnostic feature, many of the soils in these groups are characterized by reddish colours due to the presence of large amounts of iron oxide. The ferrallitic soils, ferrisols, and ferruginous tropical soils are said to be the result of relative degrees of the ferrallitization process (D'Hoore, 1964). Accordingly, the ferrallitic soils are formed where ferrallitization has been most intense, ferruginous tropical soils have formed where this process has been least intense, and ferrisols occur where the intensity of the genetic process is inter-mediate between those of the other two.

Detailed and widespread investigation of these soils in Africa, and in Tanzania in particular, has been carried out only in the last decade or two. Consequently much more work will be required before they can be classified with confidence and ease. Already workers in Tanzania have found that the grouping of these soils by the system of the *Soil Map of Africa* (D'Hoore, 1964) must be modified to accommodate the soils of that country (Anderson, 1967a; and Baker, 1970). Difficulties similar to those found by these authors are demonstrated below. The use of the Soil Taxonomy system for classifying these soils is also fraught with problems in Tanzania (Kesseba et al., 1972). Although many of these soils would fall into the oxisol order of this classification scheme, many others may be classified as alfisols or ultisols.

In the North Mkata Plain six series have been identified that may be grouped under this heading: Kingolwira, Morogoro, Vilanza, Kidunda, Kaguru, and Kidete.

Kingolwira Series

The deep, dark red to red soils of the Kingolwira series are well drained. They occur on undulating land that has slopes of up to 3 per cent. Because of the high proportion of clay the Kingolwira soils are hard during the dry season and become sticky and slightly plastic when moistened in the wet season. The areas where these soils are found do not experience regular inundation and the soils have reached an advanced stage of weathering. They can be placed in class 2 of the capability scheme.

The soils of the Kingolwira series have many features that are typical of ferrallitic soils. They have formed on clays that may be slightly sandy which have either derived from the acidic gneisses of the Uluguru and Kaguru mountains and Nguru ya Ndege or weathered in situ. The soils themselves are acidic in nature with pH values of 5.1 and less at depths below 15 centimetres. The soils at the surface have slightly higher pH values (e.g., 5.7), probably as a result of the decay of organic matter which releases bases there.

Low pH values have resulted from the intense leaching of bases from the soils during the wet season. In this acidic environment, silica becomes considerably more mobile than iron or aluminum. The silica, by removal in groundwater, 'may seep to valleys, producing 2:1 lattice clays or smectoid clays (montmorillonite), forming vertisols' (Buringh, 1970). Oxides of iron and aluminum remain, however, giving the soils their reddish colour. But in no instance does the iron in the Kingolwira soils appear as plinthite or hard concretions.

Despite the high clay content of the soils of the Kingolwira series (greater than 40 per cent in all horizons), the cation exchange capacities are very low (less than 10 me/100 g). For most ferrallitic soils, this phenomenon is explained by the presence of 1:1 lattice clays and the absence of appreciable weatherable materials (D'Hoore, 1964). This explanation is indeed valid for the Kingolwira soils as their adjusted CEC:clay ratios are less than 0.50 and they have very low silt:clay ratios in the subsoil (values of 0.03 where values less than 25 are felt to be diagnostic of ferrallitic soils).

Somewhat surprising for ferrallitic soils, however, is the fact that base saturation percentages of the Kingolwira series are relatively high — greater than 40 per cent. A possible explanation is the fact that, while these soils are well drained, heavy rainfalls occur only through a part of the year. Bases such as calcium and magnesium are therefore not entirely leached from the soils.

An apparent illuviation of clays occurs with depth, the increase in clay content from the surface to the subsoil being of the order of 10 per cent. Textural subsurface horizons of this sort are common in soils in which silica is mobile and may be moved downwards through the profile. Structural subsurface horizons do not occur in the soils of the Kingolwira series.

Although the Kingolwira soils are saturated with bases by as much as 55 per cent, this relative value does not at all reflect thier fertility status. As indicated above, these soils have cation exchange capacities of less than 10 me/100 g. The absolute amounts of bases such as calcium or potassium are extremely low and become less available with depth. Low organic matter content, C:N ratios of 9 or less, and total N percentages of 0.13 and less are very good indicators of the nitrogen deficiency of these soils. At first sight, the only plant nutrient that these soils might supply in adequate amounts is phosphorus. It must be remembered, however, that in the presence of hydrous oxides of aluminum and iron, phosphorus fixation may take place; phosphorus may also be lost temporarily to crops through the process of anion exchange where phosphorus replaces the hydroxyl groups of clay minerals, especially 1:1 lattice clays (Jackson, 1964). Measures of total P may not therefore give an indication of the amount of this nutrient that is available for plant use. The Kingolwira soils would probably require NPK fertilizer for good crop productivity.

With subsurface soils having cation exchange capacities of less than 16 me/100 g, little or no weatherable minerals, and large amounts of 1:1 type clays, features characteristic of an oxic horizon, the soils of the Kingolwira series belong to the oxisol order. They may be further classified as *typic eutrustox*. These are oxisols that are moist in some horizons for more than 3 consecutive months but are dry in some horizons for more than 60 consecutive days; they do not have indurated plinthite within 125 centimetres of the surface; and they are unmottled, unstructured, and fine-textured below the surface. Soils of the Kingolwira series have been placed in the eutrustox great group of the ustox suborder because of their high base saturation.

Morogoro Series

The Morogoro series is composed of deep, dark red sandy loams that occur on slopes of 4 per cent. Formed on sandy and silty parent materials these soils have a clay content of 10 per cent in the topsoil, rising to 14 per cent at 150 centimetres in depth. As the soils of the Morogoro series resist erosion, are rarely flooded, but are relatively coarse in texture, they have been given a capability rating of class 3.

The Morogoro soils have formed in an environment that is similar to that of the soils of the Kingolwira series. While small amounts of weatherable materials enable the subsoil to maintain a CEC of 22 me/100 g, the surface soils have low values of CEC in the order of 15 me/100 g. Since the base saturation is also low, the soils are infertile, unable to supply bases in quantities required for good crop productivity. While phosphorus is present in large quantities, the soils of the Morogoro series are nitrogen deficient, despite the fact that C:N ratios are 20 or greater. In these soils the organic matter might account for the higher cation exchange capacities than are found in the Kingolwira soils, but it does not supply nitrogen in quantity.

The soils of the Morogoro series are similar to those described from Kilosa and Tanga districts by Leutenegger, quoted by Anderson (1963). Anderson (1963) in turn classifies them as non-laterized red soils on sandy materials. Great difficulty was found in classifying these soils according to the Soil Taxonomy system. Because of their low clay content and cation exchange capacities that at depth exceed 16 me/100 g, they are not oxisols. They probably belong to the ultisol order and ustult great group because of their clay accumulation with depth, because of the low supply of bases, and because they occur in warm regions that are dry for 90 cumulative days or more in most years. Further subdivision is felt to be misleading. The Morogoro soils do have ferrallitic characteristics of low CEC and base saturations that are lower than 40 per cent.

Vilanza Series
A thin (15 centimetre) horizon of sandy clay overlies the subsurface clay of the Vilanza series soils. Despite the clay content of 40 per cent or more, these friable soils with good surface structure are well drained. The soils that comprise the Vilanza series are deep, resistant to erosion, and inundated rarely, and then only where they are found close to stream channels. As a hard cap may form at the surface because of the high clay content, they have been placed in capability class 2.

These soils are similar to the soils of the Kingolwira series in that they are formed on acid gneissic drift or gneiss that has weathered in situ. The acidic conditions of the soil (pH values of 5.8 and less) give rise to silica mobility and accumulations of iron and aluminum oxides. These soils leave very red stains on one's hands and on any other articles that come in contact with them. A textural horizon, the result of clay translocation, occurs at depths between 15 and 60 centimetres (compare the 40 per cent clay content above and below this horizon with the clay content here of 56 per cent).

These soils contain weatherable materials not found in the Kingolwira soils. A few stones occur within the soils and the silt:clay ratios are usually greater than 0.15 although they are still low (below 0.25). The cation exchange capacities of these soils are 20 me/100 g or greater, and bases are not thoroughly leached as indicated by base saturation values of approximately 50 per cent. In some instances, accumulations of carbonate concretions may be found but the amount of these accumulations is small (less than 2 per cent of the soil volume); they are found at depths of 100 centimetres or more.

In virgin conditions, the presence of bases appears adequate for the growing of most crops, but the supply would likely be rapidly depleted by cropping. For example, as much as 1.50 me/100 g of potassium may occur in the topsoil but below 60 centimetres values drop to less than 0.30 me/100 g. The Vilanza soils are extremely low in phosphorus and poorly supplied with nitrogen.

The soils of the Vilanza series are well weathered with 1:1 lattice clays dominant, but they contain quantities of 2:1 lattice clays as well. Because these soils have duller colours, higher exchange capacities, and higher base saturation values than would be normally found in ferrallitic soils, the Vilanza soils would likely be considered to be ferrisols. Using the American system, they would be classified as typic eutrustox. This, unfortunately, is the same classification given to the soils of the Kingolwira series. The USDA approach to soil classification, unlike the system used by the *Soil Map of Africa* (D'Hoore, 1964), makes no distinction between those soils having cation exchange capacities above and below 20 me/100 g (which is the case for the Vilanza and Kingolwira soils, respectively).

Kidunda Series

Popularly known as *ground-water laterites*, the ferruginous soils of the Kidunda series are imperfectly drained and have a horizon of indurated plinthite concretions beginning at approximately 60 centimetres below the surface. The concretions may be as large as 30 millimetres in diameter but are more commonly 15 millimetres in diameter. The concretionary iron is dull reddish in colour, but may show black manganese flecking when the concretions are broken open. As plant roots have great difficulty penetrating this horizon, the soils may be considered to be moderately shallow. Above this horizon, the Kidunda soils have a 20-centimetre-thick dark topsoil of sandy clay loam which is underlain by grayish sandy clay. These soils are found in areas where moderate to slightly severe sheet and gully erosion occurs if the land is cultivated. Slopes in these areas may be of the order of 2 per cent. Inundation is rare. The soils of the Kidunda series fall into class 3 of the capability scheme.

The tendency for many tropical soils to accumulate oxides of iron and aluminum is well known. In the case of the ferrallitic soils described above the sesquioxide accumulations are relative in that, in a freely draining environment, soil constituents, particularly silica, are removed by weathering and leaching processes leaving behind the iron and aluminum oxides. Although the relative accumulation of sesquioxides may have been important in the early stages of the genesis of the Kidunda soils, these soils are now being influenced by a process of absolute sesquioxide accumulation from upslope sources, as suggested by the dark grayish brown to dark gray surface horizons, the imperfect drainage, and the location of these soils on the lower portions of slopes.

Absolute accumulations of sesquioxides occur when the sesquioxides become mobile and move through the soil downslope or up and down with a fluctuating water table. In Tanzania 'aluminum normally plays a very subordinate role' (Anderson, 1963) in this process, so the following remarks are confined to iron movement and induration. In anaerobic conditions that arise with impeded drainage or with temporary waterlogging during the wet season, oxygen-requiring micro-organisms may obtain oxygen from the iron compounds in the soil. By this activity, relatively immobile ferric iron is reduced to relatively mobile ferrous iron. In the latter form, iron enters the soil solution and, as in the Kidunda soils, becomes concentrated in certain soil horizons with the alternate wetting and drying of these horizons. When the soils of the Kidunda series dry out, the ferrous iron re-oxidizes and is precipitated as ferric iron oxide concretions. Very often these concretions are found to have nuclei of small grains of quartz.

Because of the impeded drainage conditions of these soils, bases are not removed rapidly and pH levels are 6.0 and higher. Consequently, 2:1 lattice clays are present, which cause the soils to have moderately high cation exchange capacities (greater than 22 me/100 g) which are base-saturated. In the Kidunda soils the Ca + Mg: K ratios are such that potassium deficiencies do not occur. With low phosphorus and nitrogen contents, however, the soils may be considered to be of relatively low fertility.

In the terminology used for the *Soil Map of Africa* (D'Hoore, 1964), these soils would be considered as ferruginous tropical soils having impeded drainage. Using the Soil Taxonomy system, groundwater laterites may be classified under the suborder aquox, aquults, or aqualfs of the oxisol, ultisol, or alfisol order, respectively. As the soils of the Kidunda series do not have an oxic horizon characteristic of oxisols, and as base saturation values are greater than those common to ultisols, they may be considered to be alfisols. Further classification would place them as *typic tropaqualfs*. These are alfisols that occur in warm regions, are seasonally saturated with water, have base saturations greater than the 35 per cent values allowable for ultisols, have no horizons that total more than 50 centimetres with more than 35 per cent clay, do not crack when dry, and have accumulations or iron-manganese concretions.

Kaguru Series

The dark brown soils of the Kaguru series have features characteristic of impeded drainage (for example, distinct, clear mottling) despite the fact that many of them occur on slopes of 6 per cent or more. The presence of such features may be due to their fine textures, which change from clay loam to clay with depth, as well as to the fact that they are found on the lower slopes of the Kaguru-Kidete-Nguru escarpment where they receive large amounts of runoff water. In addition to being imperfectly drained, these soils suffer slightly severe to severe erosion when they are cleared for cultivation. Although they are moderately deep, these soils are placed in capability class 5 because of the cumulative limitations described above.

The soils of the Kaguru series have formed in a warm, moist environment over acidic gneiss. Despite the favourable weathering conditions, impeded drainage and the presence of significant amounts of magnesium have allowed for some

2:1 lattice clays to remain in these soils, which are therefore ferrallitized to only a small degree. Indeed, ferruginous tropical soil features, such as silt:clay ratios greater than 0.15 and high cation exchange capacities (frequently greater than 30 me/100 g) that are more than 40 per cent saturated with bases, are most common. In addition, iron concretions occur throughout the profile in amounts up to 2 per cent of the soil volume.

These base-saturated soils are considerably more fertile than the ferrallitic soils and ferrisols that have been described above. They contain large amounts of potassium (as much as 21 per cent of the exchange capacity) and are rich in phosphorus. They are poor in nitrogen, however, and have C:N ratios of 10 and less.

With such high base saturation, and because they lack an oxic horizon, the soils of the Kaguru series can be classified as *ustalfs*. These are alfisols that occur in regions having warm climates with long periods (more than 90 cumulative days) when the soil is dry. In addition they have base saturations in excess of 80 per cent. Further subdivision has not been made in the classification of these soils because they do not appear to fit any of the great groups provided in the ustalf suborder.

Kidete Series
The Kidete series is composed of moderately shallow, reddish brown sandy clay loams that have developed on gneiss, weathered in situ. These are well-drained soils but they suffer severe erosion if cleared for cultivation. They occur on rolling land where slopes of 15 per cent or more are not uncommon. With such severe limiting features, the soils of the Kidete series are rated as class 5 in the capability classification.

Appreciable amounts of sesquioxides must be present in these soils to impart the reddish colours. However, these soils are very stony and their silt:clay ratios, greater than 0.15, and adjusted CEC:clay ratios, greater than 0.50, indicate that the soils contain significant amounts of weatherable material and 2:1 lattice clays. A further indication is the cation exchange capacity, which is 31 me/100 g or more, despite the absence of large amounts of organic matter.

The cation exchange complex of the Kidete soils is base-saturated. The soils are potassium deficient, however, having Ca + Mg: K ratios below 10 centimetres that are greatly in excess (72) of the 25 considered suitable for most crops. The soils of the Kidete series are also deficient in nitrogen and phosphorus.

The classification of these ferruginous Kidete soils using the Soil Taxonomy system is as difficult as that of the soils of the Kaguru series. As ustalfs they may be considered to belong to the great group rhodustalfs because of their colours, which are redder than 5YR. But they do not have the clay content typical of these alfisols.

WEAKLY DEVELOPED SOILS ON LOOSE SEDIMENTS

The soils of this group have formed on sandy and/or clay sediments that have been laid down by fluviatile processes. The soils are immature as a consequence of the relative youth of these materials. The heterogeneity of these deposits, which show both lateral and vertical variation, is reflected in the soils. Only through very detailed survey procedures could the full range of the soil characteristics of this group be detailed or mapped in the North Mkata Plain. However, the seven soil series that are described below are felt to be typical of the most widespread soils on alluvium found in the survey area. The series have been grouped on the basis of the relative age of the sediments on which they have formed.

On Loose Sediments Not Recently Deposited

Large areas of the North Mkata Plain are covered by loose sediments which receive fresh fluviatile material only intermittently. They are inundated rarely today. However, the relative youthfulness of these deposits is still such that the soils on them are weakly developed. Three soil series are characteristic of these older sediments: Magole, Dakawa, and Kimamba.

Magole Series

The Magole series is composed of soils formed on older alluvial deposits of sandy and silty clays. In these medium-textured soils, the proportions of sand, silt, and clay vary little with depth, falling into the loam and sandy loam categories. These deep, well-drained soils suffer slight wind erosion when cleared for cultivation. As the soils of the Magole series have few physical characteristics that would limit their use for general cultivation, they have been rated as class 1 soils.

The Magole soils are relatively well supplied with bases, having cation exchange capacities greater than 20 me/100 g. These neutral to near neutral soils (pH values of 6.6 to 7.1) are base-saturated with calcium and magnesium, which make up a large proportion of the exchange capacity. Potassium, particularly in the surface soils, is sufficient for most crops, with Ca + Mg: K ratios below 25. Such moderately high exchange capacities and base saturation values are accounted for by the presence of 2:1 lattice clays (in the topsoils the adjusted CEC:clay ratios are well in excess of 0.50) and an alternating wet and dry climate which does not cause the bases to be leached from the soils. Nitrogen is available in limited amounts and would likely be depleted rapidly if the soils were heavily cropped (organic matter content is low as are the C:N ratios of 11 and less). They are probably deficient in phosphorus too, but no quantitative data are available as demonstration.

The most variable aspect of the soils of the Magole series is their colour. With depth, they change from black, to very dark brown, to black again, to dark yellowish brown, to reddish brown. These colours reflect the variable nature of the source of the materials on which the soils have formed. The reddish hues of the lower portion of the profiles are no doubt due to the fact that this material is derived from the eroded ferruginous and ferrallitic soils of the nearby, more elevated slopes.

Soils which have not been altered or have been only slightly altered by soil-forming processes are known as *entisols*. The soils of the Magole series may be labelled as *typic ustorthents*. These are entisols with textures of loamy, very fine or extremely fine sand; they have an organic matter content that decreases regularly with depth; they are not saturated with water within 150 centimetres of the surface at any time; and they occur in warm regions in which the soils are dry for 90 cumulative days or more in most years in some subhorizon.

Dakawa Series

Soils of the Dakawa series are deep, have weak structural development, and occur on older alluvial deposits that have relatively high percentages of sand and silt. Slight to moderate erosion is evident in cleared areas. Because of the erosional hazards and their coarse textures, these soils belong to capability class 2. The Dakawa soils are well drained and may be found on slopes as great as 2 per cent.

Two-to-one lattice clays are present in the soils which make up the Dakawa series. But clay percentages are low (ranging from 2 to 20 per cent and changing irregularly with depth) so cation exchange capacities are very low. A CEC of 11.55 me/100 g was measured in the upper 15 centimetres of one of these soils despite its having a clay content of only 4 per cent, while similar CEC values were calculated in horizons having three or four times as much clay. In the topsoil of the Dakawa soils, organic matter content may be as great as 7 per cent, leading to CEC values of this order.

The soils of the Dakawa series are extremely infertile. While the exchange complex is saturated with bases (68 to 80 per cent), the absolute amounts of these plant nutrients are low. Furthermore, the soils are deficient in both nitrogen and phosphorus.

Like the Magole soils, these soils are extremely variable in colour although in general appearance they are dark brown to grayish brown, changing to dark reddish brown at depths of approximately 60 centimetres. Once again, the colour variations reflect the heterogeneous mix of materials on which the soils have formed. They would be classified as *typic ustorthents*.

Kimamba Series

Coarse sands dominate the soils of the Kimamba series. Of the fine earth fraction above 50 centimetres, 78 per cent by weight is sand. The proportion increases to more than 90 per cent in the subsoil below 50 centimetres. These excessively drained soils have been forming on alluvial sand long enough, however, for a distinct surface soil to have developed. The very dark, grayish brown, granular structured, sandy loams of the upper 50 centimetres stand out strikingly from the yellowish brown, structureless, sandy materials beneath. Unless perennial crops such as sisal are grown, these soils can become eroded when cleared. They have been placed in class 3 of the capability classification.

The soils of the Kimamba series need to be fertilized for suitable crop productivity. A base saturation of slightly more than 50 per cent does not indicate that bases can be supplied in sufficient quantities, as the exchange capacities of these soils are extremely low (12 me/100 g in the topsoil, decreasing abruptly to 6 me/100 g in the subsoil). Both nitrogen and phosphorus are present in relatively low amounts.

The coarse soils of the Kimamba series have horizons that are dry for 90 cumulative days or more in most years. In addition, they are unmottled, occur in warm regions, and have base saturation values in excess of 35 per cent in horizons which occur within 125 centimetres of the surface. Accordingly, they are classified as *alfic ustipsamments* using the Soil Taxonomy system.

On Loose Sediments Recently Deposited

The soils of the second group of weakly developed soils have formed on loose sediments that have been recently deposited. The fluviatile deposits are rejuvenated annually as a result of flooding. Four soil series are characteristic: Mkundi, Mvomero, Msowero, and Wami.

Mkundi Series

The soils that comprise the Mkundi series are found on recently deposited sandy alluvium. These soils are without pedogenetic horizons, the distinct layering being due to the variable nature of the sedimentary materials. The soils are excessively drained where they occur on undulating terrain with slopes that are as great as 3 per cent but generally less. The Mkundi soils are rejuvenated with fresh alluvium annually but at the same time may be eroded easily during periods of rapid runoff and flooding. These very coarse-textured soils may be given a capability rating of class 4.

These soils are extremely infertile. Dominated as they are by coarse sands and containing little organic matter, their cation exchange capacities are very low. In the topsoil, exchange capacity values may slightly exceed 12 or 13 me/100 g, but the values diminish rapidly with depth to reach 5 me/100 g or less. In addition to low amounts of extractable bases, the Mkundi soils have little nitrogen. Phosphorus content was not determined.

While they regularly experience flooding, the soils of the Mkundi series are not waterlogged at other times. As they are immature, do not have characteristics associated with wetness, and are dry for considerable periods of time in most years, the Mkundi soils could be classed as *typic xerofluvents*.

Mvomero Series

The fine-textured soils of the Mvomero series are relatively immature, being found on recently deposited sandy clay alluvium. In these locations, with slopes up to 2 per cent, the soils may become waterlogged in the lower subsoil, apart from experiencing moderate to prolonged periods of flooding. They may be rated as class 4 capability. Because of the characteristics associated with wetness, they may be defined as *typic tropaquents*.

Adjusted CEC:clay ratios indicate that the clay fraction of these sandy clay loams has a large proportion of 2:1 lattice clays. These clays account for the moderately high exchange capacity (over 20 me/100 g) in the topsoil. With depth, however, and decreasing clay content and decreasing organic matter content, the total amount of extractable bases diminishes. Throughohout these soils, the dominant cation is calcium (usually more than 20 per cent of the exchange capacity, rising to close to 40 per cent in the subsoils).

The soils of the Mvomero series would be able to supply crops with small amounts of base nutrients. They are deficient in nitrogen and may be similarly deficient in phosphorus although quantities of this plant nutrient have not been determined.

Msowero Series

Very dark grayish brown to black colours dominate the soils of the Msowero series. These soils are poorly drained sandy clay loams that may be found on recently deposited sandy and silty clay alluvium. They occur on land that is flat, having a slope of 2 per cent or less. Very little horizonation is apparent in these soils and they are inundated by flood waters for prolonged periods annually. They fall into class 4 of the capability classification.

The soils of the Msowero series have relatively high exchange capacities that are saturated with bases, particularly calcium. Calcium is also present in the soil solution and, with a fluctuating water table, may be precipitated in the subsoil as carbonate concretions. Despite the relative availability of bases, these soils are deficient in potassium and also in nitrogen.

Another indication of the moisture regime of the Msowero soils is the mottling that occurs throughout the profile. In the upper 20 centimetres, the mottles are distinct and reddish brown in colour. Iron compounds have accumulated here, staining parts of the soil when the iron has become ferric in form with the drying out of the soils in the dry season. In the lower parts of the profile, the mottles are dark (blue-gray and black), indicating the presence of ferrous iron and perhaps manganese. The gleying of the subsoils is obvious from the chemical reduction of the iron compounds that have entered this zone.

In the Soil Taxonomy system, the soils of the Msowero series would be defined as *typic tropaquents*.

Wami Series

Found on flat terrain with little slope, the soils that comprise the Wami series experience prolonged inundation in most years. Occurring on recently deposited alluvial clays, these gray clay soils are rated in capability class 4.

For the most part these soils are structureless and massive with very little horizon differentiation. Their clay content is of the 2:1 lattice type which allows for cation exchange capacities to be moderately high (approximately 20 me/100 g or greater). Contributing to these exchange capacity values are organic matter contents of 7 and 4 per cent at the 0-15 centimetre and 15-45 centimetre depths, respectively. The exchange complexes of the Wami soils are dominated by bases. The dominant base ion is calcium, but calcium does not appear in concretionary form.

These are infertile soils, being low in potassium and nitrogen. Phosphorus may be low as well but quantitative values have not been determined.

The soils of the Wami series have features characteristic of typic tropaquents. A peculiar feature of these soils is not described by this definition, however. Because of the high clay content and significant presence of 2:1 expandable lattice clays, these soils have small cracks on occasion. A more meaningful description of the Wami soils would be provided if they were defined as vertic tropaquents. Such a subgroup is not distinguished in the Soil Taxonomy system.

HALOMORPHIC SOILS

Halomorphic soils are those that 'have been adversely modified for the growth of most crop plants by the presence or action of soluble salts' (Richards, 1954). The morphological and analytical characteristics typifying some of the soils that fall under this heading are indicated by the soils of the Mvumi series. This is the only series examined in the North Mkata Plain that has distinctive halomorphic properties throughout the profiles of the member soil individuals.

Mvumi Series

In a number of areas of the lowlands of the North Mkata Plain, soils with a coarse columnar structural and textural subsurface horizon have developed on poorly drained, sandy clay alluvium. Identified as the Mvumi series, they occur where the topography is flat and slopes do not exceed 2 per cent. Colours change with depth in these soils from dark reddish brown to dark brown and very dark grayish brown. Predominantly loamy, the soils of the Mvumi series have been given a capability rating of class 3.

In addition to the striking morphological features (notably the columnar structures in the horizons between depths of 15 and 50 centimetres), a most significant characteristic of the Mvumi soils is the high saturation with sodium ions. Except for the topsoil, this saturation constitutes 30 per cent or more of the exchange capacity and increases with depth. As the extractable sodium percentages are greater than 15 per cent, and as the sum of the ions sodium plus magnesium is greater than the sum of the ions calcium plus hydrogen, these are sodic (or alkali) soils.

Soils with sodic properties are frequently found in Tanzania on alluvial and colluvial materials at the foot of slopes and in bottomlands (Anderson, 1963). In environments that are seasonally dry but where drainage is impeded, sodium released by the natural weathering of rock minerals is inadequately leached and largely held by the exchange complex. Further, in the case of the Mvumi soils, accumulating calcium has been precipitated in the subsoils as carbonate concretions. Impeded drainage and a fluctuating water table have also permitted the accumulation of iron oxides in the form of concretions. The iron concretions frequently occur cemented together by carbonate nodules as large as 60 millimetres along their longest axis.

While the Mvumi soils are sodic, the are also saline. Although analyses were not made to determine the amount or types of soluble salts present in these soils, pH values are lower than 8.5, indicating that the Mvumi soils are saline-sodic or saline-alkali (Richards, 1954).

The soils of the Mvumi series are composed of appreciable amounts of 2:1 lattice clays, giving them cation exchange capacities greater than 20 me/100 g in the upper 50 centimetres and greater than 30 me/100 g below 50 centimetres. The exchange complexes are base-saturated (40 per cent and more) but, as indicated above, the dominant cation is sodium. Consequently, plant nutrients such as potassium make up only a small proportion of the extractable cations.

The Mvumi soils have large amounts of phosphorus but are nitrogen deficient. Crops grown on these soils would probably respond to NK fertilizer.

In the Soil Taxonomy system the soils would be found in the *alfisol order*. In addition, the soils of the Mvumi series would be classified as *typic natraqualfs* because of impeded drainage characteristics and sodium saturation values greater than 15 per cent.

LITHOSOLS

In the exploratory survey of the highland land units, it was apparent that the most widespread type of soils in these areas would be classified as *lithosols* according to the *Soil Map of Africa* (D'Hoore, 1964). In the North Mkata Plain survey area, these soils, having experienced only slight pedogenesis, have developed on the crystalline acidic rocks of the Nguru and Kaguru mountains and of Nguru ya Ndege. They occur in the Kidete land unit as well, in the vicinity of bedrock outcrops. In all of these areas, slopes are steep, and with the rapid removal of soil material through soil erosion leaving behind only a thin layer of soil over bedrock, the soils are immature.

Three soils series recognized in the North Mkata Plain survey area fall under this heading: Chala, Ndege, and Kimala. All of these lithosols would be further defined as *lithic ustipsamments* according to the Soil Taxonomy system. They are immature sandy soils that have a lithic contact within 50 centimetres of the surface. In addition, they occur in warm regions, are dry for 90 cumulative days or more in most years, and have no features characteristic of wetness.

No analytical characteristics were determined for these soils but a brief summary of their morphological features may be found below.

Chala Series
The shallow soils of the Chala series have a lithic contact at approximately 25 centimetres below the surface. These are coarse-textured soils found in highly dissected areas where slopes may be as steep as 30 per cent. The Chala soils are stony, structureless, and subject to very severe erosion. These excessively drained brown soils would be rated as class 5 in the capability classification.

Ndege Series
The soils of the Ndege series are light yellowish brown to yellowish brown in colour. They too have a lithic contact within approximately 25 centimetres of the surface. These soils are excessively drained as a result of their coarseness and their occurrence on very steep slopes. In the upper 10 centimetres, a weak granular structure may be developed. The Ndege soils are also rated as class 5.

Kimala Series

The characteristics that limit the use of these class 5 soils are similar to those described for the Chala and Ndege series: shallowness, severity of erosion, excessive natural drainage, coarseness of texture, and steep slopes. The soils of the Kimala series are brown to reddish brown in colour and may have developed granular structures in the topsoil. A lithic contact is made at approximately 30 centimetres below the surface.

5

Land Capability Summary

Fourteen mapping units have been defined to express the geographical distribution of the soils of the North Mkata Plain. These mapping units, *soil associations*, group soil series that regularly occur together in geographical proximity. Table 8 lists the soil associations and their member series that are most prominent in the land units of the survey area. An indication of the extent of the series and associations is given in the following discussion. A somewhat generalized, and considerably reduced, map of the soil associations of the North Mkata Plain (compare Pitblado, 1975) is provided in Figure 8. The specific soil features that limit the general capability of the series have been identified in Chapter 4. The capability assessments are summarized at the level of the soil mapping units in this chapter and illustrated in Figure 9.

Mkundi Association
The members of this association are the Mkundi, Mvomero, and Msowero series. Separately, these immature soils may be found along drainage channels through-out the lowlands of the survey area. They may also be found in the high-lands – particularly in the Kidete land unit – in the bottomlands of small valleys.

These series occur together in extensive, mappable (at a scale of 1:250,000) association only in the broad drainageways of the Kilosa-Turiani and Kidunda land units. Approximately 18,600 hectares, or 6.8 per cent of the survey area, have been mapped as having soils of the Mkundi association. Much of this area is associated with riverine forest vegetation or marsh, which can be identified on air photographs. In cleared areas these soils are used for rice cultivation.

TABLE 8

List of soil associations indicating their locational relationship to the land units of the North Mkata Plain

Physiographic provinces	Land units	Dominant soil associations
Highlands	Ukaguru	Chala
	Nguru	Chala
	Kidete	Kidete
	Nguru ya Ndege	Ndege
Lowlands	Kilosa-Turiani	Kaguru
		Kimamba
		Magole
		Makuyu
		Mkundi
		Mkundi-Magole
	Mkata Station-Dakawa	Kwadihombo
		Mkata
		Wami
	Kidunda	Kidunda
		Kimamba
		Magole
		Mkundi
		Mkundi-Magole
		Vilanza

The soils of the Mkundi association are limited for agriculture by prolonged periods of annual inundation in particular. In spite of this fact the most widespread soils, those of the Mkundi series, are excessively drained. In the areas where the Mvomero or Msowero series occur, the soils are poorly drained and often waterlogged even in the dry season. Using the notation outlined in Chapter 2, the Mkundi association has been labelled on the land capability map as *Bizw*. This notation results from the fact that the soil series that belong to this association fall into capability classes 3 and 4, and have limitations of the kind described above which greatly restrict their use for agriculture.

Magole Association

The Magole association is found in large areas of the northern portions of the Kilosa-Turiani and Kidunda land units. Soils of the Magole and Dakawa series cover these areas, the former being more extensive than the latter. Much of these areas is used for both large- and small-scale cultivation. Where uncleared, this

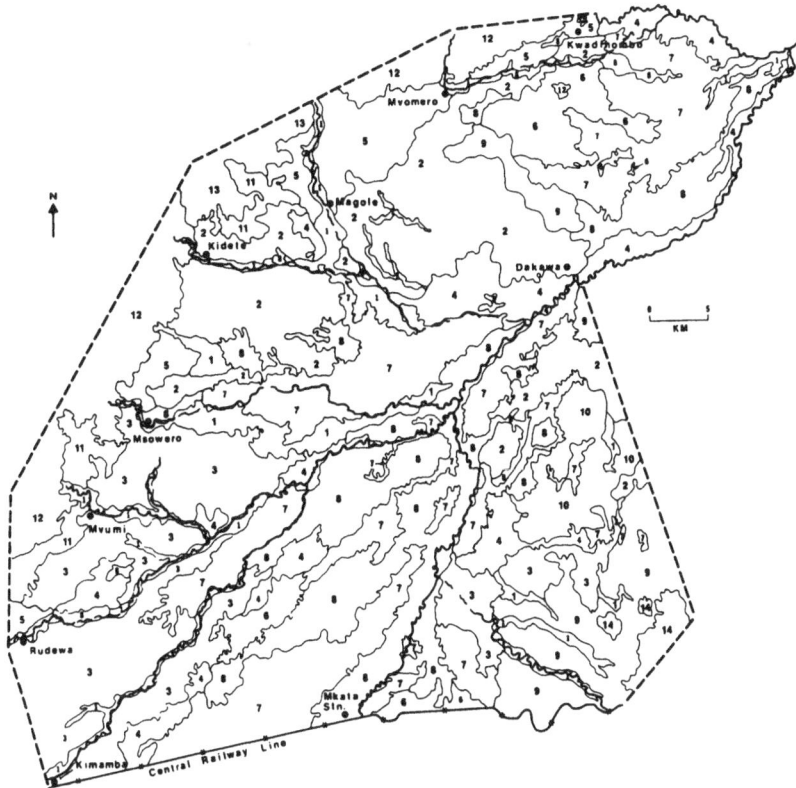

Figure 8 Soil associations of the North Mkata Plain: 1, Mkundi; 2, Magole;
3, Kimamba; 4, Mkundi-Magole; 5, Makuyu; 6, Kwadihombo; 7, Mkata; 8, Wami;
9, Vilanza; 10, Kidunda; 11, Kaguru; 12, Chala; 13, Kidete; 14, Ndege.

mapping unit is covered with dense woodland and thicket, identifiable on air
photos. The soils of this association have developed on old alluvium, and the
intricate depositional history of the parent material is reflected in the complex
distribution of the member soil series.

Soils of the Mkundi and Mvomero series occur along the minor drainageways
within the areas mapped as belonging to the Magole association. While the extent
of this subdominant series is not large, their presence may reduce the overall
capability of the Magole association somewhat.

However, the dominant series, Magole and Dakawa, have been rated as belonging to capability classes 1 and 2, respectively. Therefore the Magole association

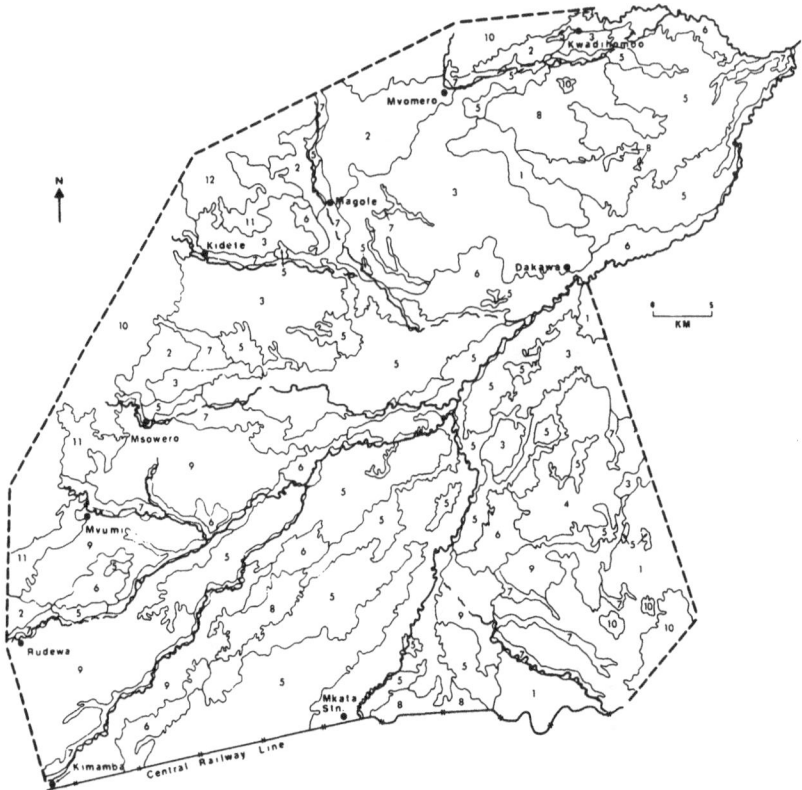

Figure 9 Land capability divisions of the North Mkata Plain: 1, At; 2, Awe; 3, Aze; 4, Bdw; 5, Bitw; 6, Biz; 7, Bizw; 8, Btwi; 9, Bzw; 10, Cdse; 11, Ce; 12, Csde. (Using soil association mapping units; see Chapter 2 for explanation of the symbols.)

has been mapped as being made up of soils that offer few or no physical limitations to cultivation — that is, capability division A. The notations z and e have been added to indicate that in some areas the water-holding capacity of the soils is limited and that soil erosion, particularly by wind action, may result from clearing and poor farm management. Slightly more than 12 per cent of the survey area, or an estimated 34,140 hectares, has been so mapped.

Kimamba Association

The Kimamba association is dominated by soils of the Kimamba and Mvumi series. These soils may be found in the southern portion of the Kilosa-Turiani land unit and the central and western areas of the Kidunda land unit. A catenary association is suggested as the Kimamba soils are found on well-drained and excessively drained sites of higher elevation, and the Mvumi soils on lower, poorly drained sites. This succession of soils ends in the drainageways with soils of the Msowero and Mkundi series. The succession does not, however, have a common parent material; the latter ranges from the older coarse sands of the Kimamba and sandy clays of the Mvumi series to the recently deposited material of variable character along the drainageways.

All of the soils that occur as members of the Kimamba association have been given capability ratings of classes 3 or 4. Their most limiting feature is either excessive natural drainage or, just the opposite, impeded drainage. Accordingly, the areas mapped as belonging to this association carry the notation *Bzw*. Despite this assessment, these areas, when properly managed, can support some specific crops such as sisal or sugarcane. In addition, this land is currently occupied in some locations by smallholders who grow crops such as maize, cotton, and millet. The Kimamba association covers slightly more than 12 per cent of the survey area, an estimated 34,400 hectares.

Mkundi-Magole Association
This mapping unit is a complex of series common to the Mkundi and Magole associations. An estimated 19,930 hectares have been mapped as such, which represent approximately 7 per cent of the survey area. This complex is found most commonly on the margins of the Kilosa-Turiani and Kidunda land units where these abut the Mkata Station-Dakawa land unit.

The soils of this association have developed on both old and recently deposited alluvium, the latter being more extensive. Soils of the Magole and Dakawa series, rated relatively highly in terms of physical capability, occur as small 'islands' distributed haphazardly throughout the mapped area. However, the general capability assessment of the Mkundi-Magole complex reflects the widespread presence of the excessively drained Mkundi soils which are flooded moreover for long periods annually — that is, it is labelled on the land capability map with the notation *Biz*. This indicates that the areas mapped as this association have soils that greatly restrict cultivation, and irrigation could be considered only after the completion of major improvement projects — particularly flood control works.

Makuyu Association
At the base of the Ukaguru-Kidete-Nguru escarpment, in the Kilosa-Turiani land unit, on sandy clay colluvium, three series occur together in catenary association. The Makuyu and Ilonga series are found on upper and middle slopes,

respectively. The imperfectly drained soils of the Kipinde series are found in sequence on the lower slopes, in close proximity to the drainageway soils of the Mkundi and Mvomero series. This soil sequence has been mapped as the Makuyu association.

As the most widespread soils that make up the Makuyu association have been given a capability rating of class 2, and as one of its member series, Ilonga, is rated as class 1, the association has been placed in capability division A. Notations w and e have been added to indicate that limitations to cultivation may arise as a result of impeded drainage or soil erosion. Otherwise these soils have physical characteristics that are favourable to sustained cultivation and, given proper management, they are suitable for irrigation.

At present most of the soils of the Makuyu association are under some form of cultivation – either by the mixed cropping systems of smallholders or the monoculture practices of large-scale enterprises. In extent, these soils cover an estimated 11,480 hectares or slightly more than 4 per cent of the survey area.

Kwadihombo Association

The brown vertisols of the Kwadihombo series dominate the Kwadihombo association. The fine-textured, poorly drained soils are inundated infrequently and then for short periods only. They have been given a capability rating of class 3. They are associated with other fine-textured soils, particularly those of the Mkata series, which experience prolonged annual inundation. Accordingly, the soils of the Kwadihombo association have limitations that greatly restrict cultivation, and the association has been labelled with the notation Btwi.

Throughout the survey area the soils of this association support a semi-natural vegetation cover of wooded grassland. In comparison with other wooded grassland areas in the survey area, the woody species here are closely spaced. Less dense wooded grassland is more typical of the areas dominated by the soils of the Wami series (see Wami association below) which make up a small proportion of this association. These differences can be observed on air photos, and provide an aid in mapping the associations. In addition, grassland areas here generally indicate soils of the Mkata series, and riverine forest is a fair indicator of the presence of soils of the Mkundi, Msowero, and Mvomero series.

The soils of the Kwadihombo association are not cultivated except for the very small areas of smallholder cultivation east of Mkata Station and south of the village of Kwadihombo. They are primarily used for periodic grazing by cattle of the Baraguyu. At some sites, moderate to slightly severe erosion may occur as the result of overgrazing and trampling by the cattle. This association covers slightly more than 4 per cent of the survey area, or 11,600 hectares.

Mkata Association

The dominant soils of the Mkata association are the Mkata, Kwavi, and Wami series. For the most part, this association can be considered as typical of the *mbugas* (frequently inundated flat lands) that are common in Tanzania. The most widespread of the series, the Mkata and Kwavi, support grassland vegetation although wooded grassland may appear on these soils on occasion. Wooded grassland, with woody species widely separated, also appears where the Wami soils occur. Subdominant series in the areas mapped as belonging to the Mkata association may include Mkundi, Msowero, and Mvomero. Localized areas near the Central Railway Line and towards Kimamba may have the saline-sodic soils of the Mvumi series.

The Mkata association is the most widespread association of soils in both the Mkata Station-Dakawa land unit and the whole North Mkata Plain survey area. It covers 18 per cent of the latter, or slightly more than 43,300 hectares. The soils are for the most part inundated for prolonged periods annually. They also have very fine textures and are poorly drained. As all the member series have a capability rating of classes 3 and especially 4, the Mkata association has a general capability assessment of *Bitw*.

Very little cultivation is carried out on these soils. They are used primarily for the grazing of cattle by the Baraguyu and by the government cattle ranch which extends north from the Central Railway Line to the Wami River. Without major improvement projects, which would include flood control and land drainage, these soils are unsuitable for sustained, widespread cultivation.

Wami Association

The second largest in extent, the Wami association covers an estimated 35,910 hectares, or 13 per cent of the survey area. In order, the most widespread series that make up this association are the Wami, Msowero, and Mkata. Subdominant series include the Kwavi and Mvomero. As the majority of these soils present severe limitations to agricultural use because of prolonged annual flooding, fine texture, and poor natural drainage, the association has been mapped with the notation *Bitw*.

The soils of this association occur principally in the Mkata Station-Dakawa land unit, usually in close proximity to the Wami and Mkata rivers or their major tributaries. Here they support open wooded grassland, thorn thicket, and grassland. Except for the few hectares near Mkata Station, these soils are not cultivated. As in the case of the two associations just described, the Wami association soils are used for periodic grazing. In the small areas where this association occurs in the Kilosa-Turiani land unit, the soils may be used for the cultivation of rice and sugarcane.

Vilanza Association

Four of the ferrallitized soils of the North Mkata Plain may occur in a catenary association, which has been named Vilanza. In sequence downslope, they are the Vilanza, Kingolwira, Morogoro, and Kidunda series. At the base of this succession, in bottomlands, the soils may be characteristic of either the Mkundi or Mkata series.

The Vilanza association encompasses areas in the Kidunda and Kilosa-Turiani land units; in the latter the Kidunda member of this succession is usually missing. The three remaining well-drained members, two of which are individually given a capability rating of class 2, are widespread and allow for a general capability assessment of the Vilanza association of *At*. Only the fine texture, which causes some capping of these soils when they dry out, would constitute a physical limitation for general agricultural use.

Unfortunately, most of the areas mapped as belonging to the Vilanza association are found where rainfall is limited for long periods of time and where surface drainage waters are not in adequate supply for irrigation. At present much of this mapping unit has been set aside as a forest or fuel reserve. For these reasons the Vilanza association soils are rarely used for cultivation. In addition, very little use is made of these areas for grazing, as the woodlands and thickets which the soils support are infested with tsetse fly.

At total of 17,340 hectares, slightly more than 6 per cent of the survey area, has been mapped as the Vilanza association.

Kidunda Association

The Kidunda association includes areas in the Kidunda land unit where the Kidunda member of the catenary association mentioned above is not only present but widespread. In fact, the catenary member often missing is the Vilanza series. The 8,480 hectares (approximately 3 per cent of the survey area) so mapped have a general capability rating of *Bdw*. This rating reflects the capability class 3 assessment of the Kidunda and Morogoro series. It is most particularly indicative of the limitations of depth and drainage, and the associated iron induration characteristic of the Kidunda series.

Within the areas mapped as the Kidunda association, the soils support natural vegetation of woodland and wooded bushland. Little of these areas has been cleared for cultivation, and grazing is limited because of the lack of any surface waters and the presence of tsetse fly.

Kaguru Association

At the base of the Ukaguru-Kidete-Nguru escarpment in the Kilosa-Turiani land unit is an area of very severely eroded soils. West of the Kilosa-Turiani road between Msowero and Rudewa and north of Kidete, the soils appear to be eroded phases of the Kaguru series or of the member series of the Vilanza association. As very little survey time was spent in this area, it is difficult to suggest a cause for the widespread erosion. However, little use will likely be made of these soils for a long time to come and the area has been mapped with a general capability notation of *Ce*. The area covers 6,750 hectares, or less than 3 per cent of the survey area.

Chala Association

The Chala association is coincident with the Kaguru and Nguru land units. As indicated in Chapter 2, little survey time was alloted to these land units and only three series have been identified in a section that covers slightly more than 6 per cent of the study area, or an estimated 17,540 hectares.

The three series are Chala, Kimala, and Kaguru, the Chala series being the most widespread. All of the soils present limitations – particularly shallow depth, steep slopes, and very severe erosion – that militate against either arable or pastoral use. The Chala mapping unit has accordingly been given the capability notation of *Cdse*.

Kidete Association

In the areas mapped as the Kidete association, the most widespread soils belong to the Kidete series. These moderately shallow soils are well drained, but they occur on slopes of up to 15 per cent and should therefore not be used for agricultural or pastoral activities. In the past, the soils have been cleared for cultivation or for firewood and as a result are now severely eroded.

Commonly associated with these soils are the thin lithosols of the Chala and Kimala series. All three of these series have been rated as capability class 5. Accordingly, the Kidete association has been mapped with the notation *Cdse*.

Other soils that occur within this mapping unit have characteristics similar to those of the immature, alluvial soils of the Mkundi association – especially the Mkundi series. These soils occur in the bottomlands of the small, V-shaped valleys of the Kidete land unit.

The Kidete association makes up only a small proportion of the survey area – slightly less than 2 per cent of the area, or 5,240 hectares.

Ndege Association

The boundaries of the Ndege association outline the portion of Nguru ya Ndege that is found within the Wami River Basin. The mapping unit also includes some large bedrock outcrops that lie immediately to the north and west of Nguru ya Ndege. Little soil has developed in these areas. When soil has formed, as in the

TABLE 9

Extent of the soil associations of the North Mkata Plain by dominant soil types

Dominant soil types	Soil associations	Area	
		Hectares	Percentage
Vertisols	Kwadihombo	11,600	4.3
	Mkata	49,350	18.0
		60,950	22.3
Eutrophic Brown Soils	Makuyu	11,480	4.2
Weakly Developed Soils:			
(a) On loose sediments	Kimamba	34,400	12.6
not recently deposited	Magole	34,140	12.4
		68,540	25.0
(b) On loose sediments	Mkundi	18,600	6.8
recently deposited	Mkundi-Magole	19,930	7.3
	Wami	35,910	13.1
		74,440	27.2
Ferrallitic soils, Ferrisols,	Kaguru	6,750	2.5
and Ferruginous tropical soils	Kidunda	8,840	3.2
	Vilanza	17,340	6.3
		32,930	12.0
Lithosols	Chala	17,540	6.4
	Kidete	5,240	1.9
	Ndege	2,740	1.0
		25,520	9.3
TOTAL		273,860	100.0

case of the Ndege series, it is found on steep slopes and is shallow, stony, and severely eroded. The notation *Cdse* signifies that these soils are entirely unsuitable for any arable or pastoral activity.

SUMMARY

Table 9 summarizes the geographical extent of the soil associations of the North Mkata Plain by grouping the mapping units under headings that indicate the

TABLE 10

Extent of the soil associations of the North Mkata Plain by capability division

Capability division	Soil associations	Area	
		Hectares	Percentage
A	Magole	62,960	23.0
	Makuyu		
	Vilanza		
B	Kidunda	178,270	65.2
	Kimamba		
	Kwadihombo		
	Mkata		
	Mkundi		
	Mkundi-Magole		
	Wami		
C	Chala	32,270	11.8
	Kaguru		
	Kidete		
	Ndege		

predominant types of soil making up an association. These headings are similar to those found in the legend of the *Soil Map of Africa* (D'Hoore, 1964).

As the table shows, the majority of the soils of the survey area are relatively immature. These soils (Weakly Developed Soils on Loose Sediments and Lithosols) cover slightly more than 61 per cent of the study area, or an estimated 168,500 hectares. The remainder of the survey area is made up of more mature soils, especially the Vertisols, which cover some 22 per cent of the area, approximately 60,950 hectares.

In terms of land capability, very little of the North Mkata Plain provides an environment in which the physical characteristics of the soils are eminently suitable for cultivation. Only two of the twenty-three soil series have been rated as capability class 1. However, an additional five series have only moderate limitations to general arable use, allowing for some areas, where these series occur together, to be mapped as capability division A. In the North Mkata Plain, mapping units that can be given this general capability assessment cover an estimated 62,960 hectares, approximately 23 per cent of the survey area.

Unfortunately, these figures are somewhat misleading. As indicated in Chapter 4, the majority of the soils in these capability division A mapping units are chemically infertile. Provided that fertilizer applications are made where needed, this problem can be overcome, and on this basis, these areas could sustain intensive agricultural development. But a more limiting environmental problem is the fact that many of the physically capable soils are found in area that experience long periods of drought. In addition, mapping units such as the Vilanza association occur where the supply of water is not adequate for irrigation agriculture.

As Table 10 shows, by far the most extensive soils of the study area have limitations that greatly restrict their potential for general arable use. Many of these soils might be made suitable by the provision of flood control works and with the assistance of artificial drainage. But these projects would be extremely expensive and beyond the resources of the local inhabitants or of the Tanzanian government. Without projects such as these, however, most of these soils will not be developed intensively for agriculture in the near future.

The remainder of the survey area, as shown in Table 10, has such severe limitations that both arable and pastoral activities should be avoided. Capability division C areas have soils that are severely eroded and have steep slopes. The land here should be left for wildlife and/or forestry activities. One of the areas, Nguru ya Ndege, has already been set aside as a forest reserve; the other areas should be designated as reserves as well.

PART III

Land Tenure and Land Use

6

North Mkata Plain Overview

In the North Mkata Plain the related land tenure and land use patterns have originated and evolved from a historical admixture of traditional and modern systems of agriculture, elements of which are common throughout much of Tanzania. This admixture results from two colonial periods since the late 1800s and the coming of Independence in 1961. For this reason it is useful to examine the development of some land tenure characteristics and selected land laws of Tanzania prior to discussing the North Mkata systems.

THE NATIONAL FRAMEWORK

Prior to Tanzania's colonial periods land was held according to native custom and law. As there is little documentary evidence detailing the land tenure characteristics of this period, it is difficult to reconstruct the patterns except through the interpolation of twentieth-century studies. But these pieces of evidence record only remnants of tenure systems – systems that are continually evolving under changing physical and socio-economic pressures.

Another factor that militates against a useful reconstruction of a traditional Tanzanian land tenure system is the fact that there is no single dominant national system. With over 120 recognized tribal groupings, and with the numerous intra-tribal variations, it may be fruitless to contemplate a *typical* traditional system for Tanzania. Despite the summaries that do exist, it would appear far more profitable to examine individual tribal or intra-tribal systems.

It is felt, therefore, that for the purpose of this book traditional land tenure should be examined with specific reference to the North Mkata Plain. Like many Tanzanian land tenure systems, those of the survey area are characterized by the holding of land by a social group – a tribal or clan group. Rights to occupancy and to various occupancy practices are gained by individuals on the basis of their membership in the social group or by permission of the elders of the group in whom land allocation powers have been vested. A more detailed examination of a traditional land tenure system is provided in the next chapter, where the land tenure and land use systems of the Ngulu are discussed.

European interest in the ownership and holding of land in Tanzania began with German intervention in the 1880s (see Iliffe, 1969; Gwassa, 1969). With the treaties that he had gathered in 1884 and the subsequent German Imperial Charter of 1885, Dr Karl Peters formed the German East Africa Company in 1887. Intent on economic exploitation, this company and later subcompanies collected additional treaties from Tanzanian rulers, who unknowingly surrendered their land and authority to the Germans. In late 1888, the German East Africa Company and its affiliates had established plantations on these treaty lands along the Indian Ocean coast and in the Pangani and Umba river basins. By 1891, when the Imperial German Government took over the administration of Tanganyika, settlements and plantations had been started at such centres as Tanga, Mpwapwa, Tabora, and Ujiji and in the Usamabara, Kilimanjaro, and Meru areas. After a number of failures due to inadequate knowledge of the techniques required for tropical agriculture, and on some plantations the lack of adequate African labour, the German government encouraged the settlement of highland areas along European farming lines. 'Land grabbing by settlers followed in areas of high potential' (Gwassa, 1969). To control the acquisition of land and to further its development, the German government enacted a series of land ordinances and regulations. Two of these documents are examined here.

The German Imperial Decree of 26 November 1895 was enacted to deal with the acquisition and conveyancing of both Crown lands and lands in general in German East Africa. Some of its important articles are reproduced below (Anon., 1916):

Excepting where ownership can be shown by private or judicial persons ... all land in German East Africa shall be regarded as unowned. Ownership to such land is vested in the Empire.

When land is taken as Crown land, in localities where there are existing native settlements, areas must be reserved sufficient for native cultivation or other use, regard being had to future increase of population.

The transfer of Crown land is effected through the Governor either by conveyance of ownership or lease.

The transfer of ownership or lease of Township lands of more than one hectare in extent, and of all other lands, by natives to non-natives for a period exceeding fifteen years, may not take place without the consent of the Governor.

By these and other provisions of the decree, the German government hoped to control the acquisition and conveyancing of land in Tanganyika. Existing alienated freehold land was recognized but further freehold alienation of land could only be acquired through application. Native rights in the holding and use of land were not protected in any adequate way. The decree and later attempts at land registration did little to stop landgrabbing for in pratice most applications for freehold land were granted.

But of more concern to the government was that the land once acquired, often remained undeveloped. To alter the situation, it was made more difficult to obtain freehold land. From 1903 large tracts of Crown land, rather than being sold immediately, were leased for the purpose of cultivation for a period of up to twenty-five years under the terms of the Circular Regarding Lease of Crown Lands. A moderate rent was demanded, but, what is more significant, development conditions were applied to the lease: 'The lessee is bound within one year after the conclusion of the lease to develop the land by cultivation or other means and so to continue therein, that yearly, at least, one-tenth of the land leased shall be brought under cultivation or otherwise utilised' (German East Africa, 1903). When half of the leased land was brought under development according to these regulations, the lessee had the option of purchasing the land outright or continuing to hold the land according to the terms of the existing lease.

By the end of the German colonial period, land in Tanganyika was either Crown land or land held by independent title. Freehold titles were obtained either under the provisions made by the Imperial Decree, 1895 and related legislation, or by the conversion of leasehold estates into fee-simple estates. About 1 per cent of the land of Tanganyika was so held. Development conditions were applied to the land held as leasehold estates. Except where freehold or leasehold tenure pertained, native law and custom regarding land were little altered even though the land was technically Crown land.

Tanganyika became a trust territory of the League of Nations after the First World War and fell under British administration according to the articles of the British Mandate for East Africa. Because of the obligations imposed by the mandate, and later the UN Trusteeship Agreement for Tanganyika, native rights in land were protected.

On 26 January 1923 the Land Ordinance, 1923 was enacted by the Government of Tanganyika. This had the potential to alter native land tenure, for it declared that the whole of the lands of the territory, whether occupied or unoccupied on the date of the commencement of the ordinance, were public lands, 'provided that the validity of any title to land or any interest therein lawfully acquired before the commencement of the Ordinance was not affected.'

By this ordinance, therefore, the governor of the territory was permitted to alienate public land by grant of *right of occupancy* for a term not exceeding ninety-nine years. But rights of occupancy could be granted to both Africans and non-Africans. In practice, two types of right to occupancy arose. The first could be acquired by a direct grant from the governor to an African or a non-African. The second was acquired in recognition of the title of a native or a native community lawfully using or occupying land in accordance with native law and custom.

Technically, therefore, most Africans in Tanganyika during the British period held their land under right of occupancy arising out of African customary law. Customary laws of inheritance governed succession, and the occupier paid no rent. Owners holding land by grant, however, tendered rent. On the death of the occupant, inheritance was governed by the rules of statutory law. Both forms of right of occupancy were subject to development conditions. However, these conditions were rarely applied to rights of occupancy arising out of African customary law, although attempts were made to encourage better land use techniques. These conditions were applied, however, to rights of occupancy by grant, and occupancy could be terminated for different reasons, such as: non-payment of rent, taxes, or other dues imposed upon the land; abandonment or non-use of land for a period of five years; a breach of any of the conditions contained in a certificate of occupancy; or a breach of any regulations under the Land Ordinance relating to the transfer or other dealing with rights of occupancy. Another difference between the two types of occupancy was that rights of occupancy by grant were to be registered under the Land Registry Ordinance, 1923 and subsequent amendments and substitutions, such as the Registration of Documents Ordinance, 1924 and the Land Registration Ordinance, 1954, while land held by customary law was not registered.

An important aspect of the Land Ordinance, 1923 was that in recognizing the validity of any title to land acquired before 1923, it in effect recognized the freehold tenure of land alienated in the German period. Thus, although no further freehold alienation was provided for by the ordinance, the existing freehold land was allowed to remain.

Perhaps the most notable aspect of the Land Ordinance, 1923 is its emphasis on the use and development of land. Very little importance is laid on security of tenure. In many ways this foreshadows the present-day land tenure policies of Tanzania, for 'this philosophy is clearly reflected in the post-Independence land law legislation which aims at control of land use' (McAuslan, 1967).

The year 1923 was also significant for the development of land law in Tanzania because of the enactment of the Land (Law of Property and Conveyancing) Ordinance, 1923. This ordinance provided that the 'law relating to real and personal property, mortgagor and mortgagee, landlord and tenant, and trusts and trustees in force in England on the first day of January, 1922, shall apply to real and personal property, mortgages, leases, and tenancies, and trusts and trustees in Tanganyika in like manner as it applies in England, and the English law and practice of conveyancing in force in England on the day aforesaid shall be in force in Tanganyika.' In practice, this law applied to only a small part of Tanzania because, by its provisions, it did not affect land held under native law and custom. Even so, it was a clumsy attempt to transfer European law to Africa and its ill effects are still being felt. 'This out-dated law based on policies which have never been, nor will ever be, part of tenure policies of Tanganyika is a definite hindrance to the evolution of a modern legal regime which adequately reflects the policies of the present government' (McAuslan, 1967).

Further development conditions were imposed on land held by grant by the Land Regulations, 1926 and the Land (Pastoral Purposes) Regulations, 1927. These laws were enacted to encourage the investment of specific sums of money for specific improvements within specified periods of time. Unfortunately, the loopholes that could be found in these regulations rendered them ineffective in terms of land development. For example, under the Land Regulations, 1926, it was open to the occupier to select the improvements to be undertaken from a list of improvements provided. Many spent the required sum of money demanded by the regulations on residential buildings rather than on developing the land. In many cases, the government did not have adequate powers to compel an occupier to make land productive. As a consequence, even when the land was granted specifically for agricultural purposes, it remained undeveloped.

The Land Regulations, 1948 were enacted to change this situation by adding development conditions to the regulations of 1926. Land was divided according to its predominant use, and development conditions were set. In addition, a right of occupancy could not be disposed of without the prior permission of the land authorities. In practice, permission was not granted unless the development requirements under the terms of the right of occupancy had been fulfilled. In this way, speculation in land was prevented and pressure was applied on an occupier to complete the development conditions.

Following the viewpoints of the East Africa Royal Commission (1955), the Government of Tanganyika recommended (Tanganyika, 1958):

The provision by statute of a form of tenure which is individual, exclusive, secure, unlimited in time and negotiable – in other words, individual ownership of land, to be called *freehold*.

The encouragement of the transition from native customary tenure into *freehold* in appropriate areas.

It was felt that significant land tenure changes of the sort proposed in 1955 and 1958 would facilitate the development of cash-cropping and the introduction of modern farming methods, as well as reduce the deterioration of soil fertility and the number of land disputes that were apparently being caused by a significant increase in the African population.

However, no tenure rules of this type were enacted as the proposals were soundly condemned, particularly by the political party Tanganyika African National Union (TANU), which at that time was gathering strength. As Nyerere (1958) wrote, 'I am opposed to the proposed Government solution ... If we allow land to be sold like a robe, within a short period there would only be a few Africans possessing land in Tanganyika and all the others would be tenants.'

By the end of the British colonial period, three types of agricultural land tenure had been defined and adopted: freehold arising out of German alienation, rights of occupancy arising out of African customary law, and rights of occupancy by grant. Land tenure and land use were closely connected, with development conditions being attached to the rights to hold land, especially land held by right of occupancy by grant. Technically, development conditions also applied to rights of occupancy arising out of African Law, but customary land law was little altered directly.

Since Independence, government legislation regarding land tenure in Tanzania has continued to stress the importance of land use regulations. In addition, a great deal of emphasis has been placed on the role of administrators and administrative bodies in interpreting law.

The government paper *Land Tenure Reform Proposals* (Tanganyika, 1962) foreshadowed some of the major land law enactments of the 1960s. Its preamble reads:

The independent Government of Tanganyika has left no room for doubt about the need for strengthening the economy of Tanganyika as the means for raising the standard of living of the people. Having regard for the importance of agriculture in the national economy at present, Government has already taken steps towards procuring the development of land to the greatest possible extent and with the greatest possible speed. In particular, it has vigorously urged all Africans occupying land under native law and custom to develop their land to the full and where practicable, to expand their holdings. In consonance with this campaign, Government announced in the National Assembly its decision to convert freehold titles to leasehold. Since the date of that announcement, Government has come to the conclusion that some land held on rights of occupancy issued before the Land Regulations, 1948 were applied, is inadequately developed. Government has therefore decided to take steps to procure the development of such land also.

The immediate response to these proposals was the enactment of the Freehold Titles (Conversion) and Government Leases Act, 1963 and the Rights of Occupancy (Development Conditions) Act, 1963. The latter enactment provides that rights of occupancy granted before December 1948 shall have the development conditions of the Land Regulations, 1948. Where the occupier felt that these development conditions were too heavy, he could apply to the commissioner of lands to have conditions lifted. The commissioner was empowered to resolve the conflict between the express and the implied conditions. An agricultural land tribunal could be set up to which appeals of the commissioner's decisions could be made. The decision of the tribunal was final and not subject to review in any court. It is obvious from the above that it was intended to increase the power over the control of land use, and also to place this power in the hands of administrators and administrative bodies. Administrative power was increased not only through the provisions for the agricultural land tribunal, but through the other provisions of the act as well.

The other act which was derived directly from the Land Tenure Reform Proposals, 1962 was the Freehold Titles (Conversion) and the Government Leases Act, 1963. Basically, the main objective of this act, as with the act just discussed, was the proper development of land. It provided for the conversion of estates in fee-simple into government leasehold estates for a term of ninety-nine years, and the application of development conditions to these and other government leaseholds. With respect to development conditions, the country was divided into: (i) urban holdings in planning areas, (ii) urban holdings in other areas, (iii) large rural holdings, and (iv) small rural holdings. Large rural holdings were defined as 'leased land, not being an urban holding, exceeding one hundred and twenty acres [50 hectares] in area.' The development requirements for these holdings are outlined in the Land Regulations, 1948, and the amendments to this act found in the Land (Amendment) Regulations, 1958. A small rural holding was defined as 'leased land, not being an urban holding, of one hundred and twenty acres [50 hectares] or less.' When a small rural holding was declared ripe for development, the commissioner of lands was permitted to annex such development requirements as were felt necessary.

Similar to the Rights of Occupancy (Development Conditions) Act, 1963, the Freehold Titles (Conversion) and, Government Leases Act, 1963 provided for appeals against all or some of the annexed development conditions through a land tribunal. The tribunal could alter, vary, or revoke development conditions and other decisions made by the commissioner of lands under the act.

President Nyerere was elected to office in 1962. In his Inaugural Address (Nyerere, 1962a) he foresaw what was to become the Villagization and Settlement Program with his pronouncement:

The hand-hoe will not bring us the things we need today ... We have got to begin using the plough and the tractor instead. But our people do not have enough money, and nor has the Government, to provide each family with a tractor. So what we must do is to try and make it possible for groups of farmers to get together and share the cost of the use of a tractor between them. But we cannot even do this if our people are going to continue living scattered over a wide area ... The first and absolutely essential thing to do ... is to begin living in proper villages ... unless we do we shall not be able to provide ourselves with the things we need to develop our land and to raise our standard of living.

To effect the villagization program and to set up some sixty-nine model village settlement schemes, a number of acts and amendments to existing legislation were adopted. These included the Public Land (Preserved Areas) Ordinance (Amendment), 1965, the Land Tenure (Village Settlements) Act, 1965, and the Rural Settlement Commission Act, 1965. But the program was not the success intended. Fraser-Smith (1965), Newiger (1968), Heijnen (1969), and others have examined the reasons for the program's failure. The reasons include: over-capitalization of the schemes; the failure to provide adequate training for managers and technical staff; the failure to screen potential settlers, with the result that many of the settlers, the urban unemployed, lacked agricultural know-how; and the failure to provide any degree of security of tenure. However, the legislation did allow for the partial implementation of the policy of *ujamaa vijijini* (co-operative or socialistic villages) that was introduced in 1967 to replace the Villagization and Settlement Program (Nyerere, 1967).

One of the objectives of developing ujamaa villages was to obtain the economic advantages of large-scale farming through the introduction of modern agricultural techniques and by farming village land collectively. This policy evolved from the improvement and transformation approaches which were incorporated in the Tanganyika Five-Year Plan (Tanganyika, 1964) and given new direction in the frontal approach of the Tanzania Second Five-Year Plan (Tanzania, 1969a).

With the introduction of the policy of ujamaa villages, the government had to face the problem of land rights already existing under customary law. The amendment in 1965 of the Public Land (Preserved Areas) Ordinance, 1954 partly met this problem. The effect was to extinguish customary rights in the settlement areas. But it did not address the problem of clearly defining the land tenure position in ujamaa villages. The tenure problem for ujamaa villages was not that of acquisition, as this was provided for by the legal instrument of the Land Acquisition Act, 1967, which updated the Land Acquisition Ordinance, 1926. Rather, the problem was that of finding a suitable replacement for the tenure systems that were extinguished.

The Government Leaseholds (Conversion to Rights of Occupancy) Act, 1969 reduced all leasehold estates to rights of occupancy. By it the Freehold Titles (Conversion) and Government Leases Act, 1963 was disapplied. Significantly, all land held under the 1979 act became subject to the development conditions of the Land Ordinance, another attempt to foster agricultural growth through legislation. The imposition of development conditions was further strengthened by the Land Laws (Miscellaneous Amendments) Act, 1970, the Land Ordinance (Amendment) Act, 1974, and the Land (Rent and Service Charge) Act, 1974.

The problems of implementing the ujamaa villages policy and the shift to a villagization policy de-emphasizing the ujamaa philosophy have been examined by Ellman (1970), Rald (1970), Sandberg (1974), Boesen (1976), and others. Notwithstanding the agreements against it, the legislation pertaining to this shift in policy did allow for a more concrete definition of land tenure rules and regulations within new villages, whether ujamaa or not.

The Villages and Ujamaa Villages (Registration, Designation and Administration) Act, 1975 and the Villages and Ujamaa Villages Regulations, 1975 provided such settlements to be registered as grants of occupancy. The acts and subsequent government notices enumerated the rights of village members to the use of land for individual purposes and the responsibilities of village councils in land transfers and allocations and land use planning.

Today in Tanzania, the number of forms of agricultural land tenure has effectively been reduced to two: rights of occupancy arising out of African customary law and rights of occupancy by grant. Emphasis on the proper management and use of land continues to be written into law. These tenure forms and their related land use patterns in the North Mkata Plain can be readily observed.

THE NORTH MKATA PLAIN

On the basis of the field survey of 1970–71, the areas of the North Mkata Plain held under the two forms of land tenure (occupancy by grant and by customary law) have been divided according to dominant rural activity. These land uses are summarized in Table 11 and are mapped in Figure 10. The corresponding modes of occupancy and associated uses of land are discussed in the remainder of this chapter.

Rights of Occupancy by Grant
In the North Mkata Plain slightly less than half the land (approximately 131,800 hectares) is held by rights of occupancy by grant. The rules and regulations governing the holding, transfer, and use of this land are embodied in Tanzanian statutory law. In the study area, lands held in this manner can be discussed under seven headings.

TABLE 11

North Mkata Plain: Land tenure and land use

	Area	
	Hectares	Percentage
I Rights of occupancy by grant		
Sisal estates	29,500	10.8
Wami Irrigation Scheme	7,300	2.7
Wami Prison Farm	3,500	1.3
Nguru Forest Reserve	32,900	12.0
NDC Cattle Ranch	43,300	15.8
Ujamaa villages*	—	—
Other largeholdings	15,300	5.6
	131,800	48.2
II Customary rights of occupancy		
Recently cultivated smallholdings	31,440	11.4
Unoccupied or periodically		
grazed land	94,420	34.5
Wakwavi Settlement Scheme	16,200	5.9
	142,060	51.8
TOTAL	273,860	100.0

* As indicated in the text, the area covered by ujamaa villages at the time of the survey was minimal. It would have made up less than 0.01 per cent in this table.

Sisal Estates

European influence in the agriculture of the North Mkata Plain began with the planting of cotton by German and Greek settlers in the Kilosa-Kimamba area in the early 1900s. However, by 1909, the Central Railway Line from Dar es Salaam had been extended to Kilosa. With this extension for transportation of farm produce, the settlers found sisal to be a far more profitable crop and began to expand the number and size of sisal estates in the area. The estates at this time were held under the land alienation regulations of the German Imperial Decree of 1895 and the *Circular* of 1903. The most active period of expansion of sisal acreage came in the 1930s. In this period sisal estates extended northeast along the Kilosa-Turiani road to Kwadihombo, and one estate was located southeast of Dakawa on the Morogoro-Mvomero road. By this time, land alienation was subject to the rules of the Land Regulations, 1926. Although it is difficult to determine, it would appear that the application and enforcement of the regulations had little effect on the land use patterns of the period. As this was an era of expansion, the necessity of enforcement was likely to have been irrelevant.

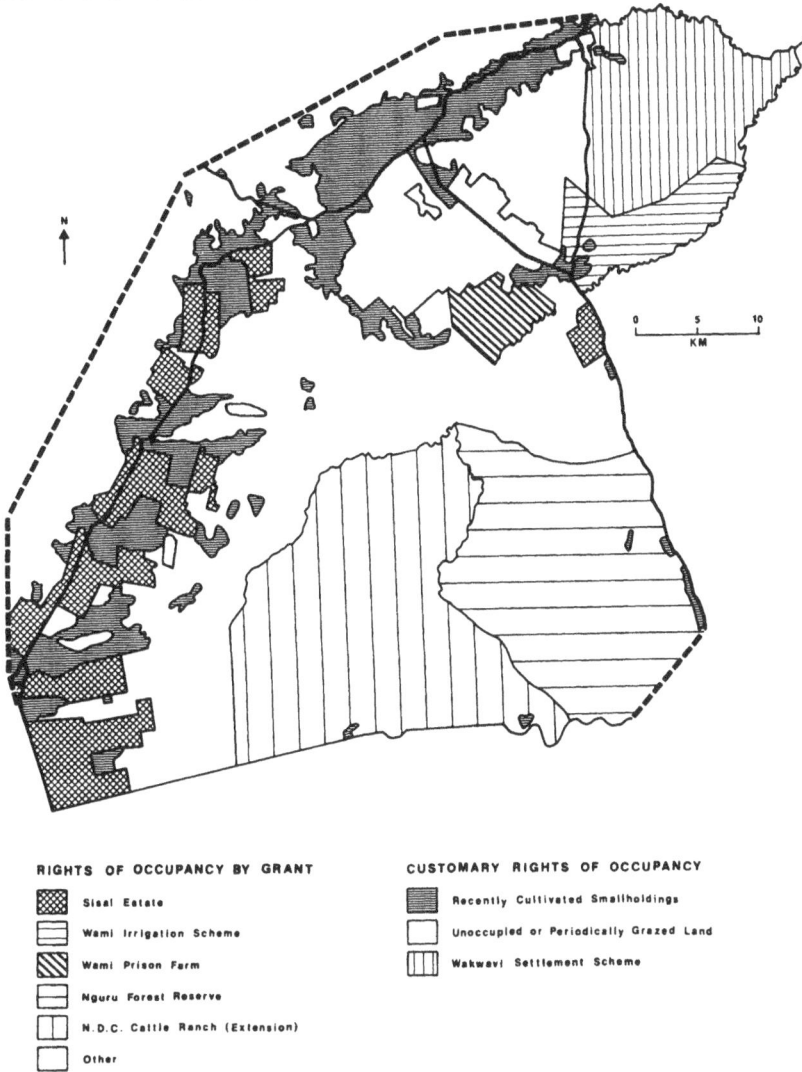

RIGHTS OF OCCUPANCY BY GRANT

- Sisal Estate
- Wami Irrigation Scheme
- Wami Prison Farm
- Nguru Forest Reserve
- N.D.C. Cattle Ranch (Extension)
- Other

CUSTOMARY RIGHTS OF OCCUPANCY

- Recently Cultivated Smallholdings
- Unoccupied or Periodically Grazed Land
- Wakwavi Settlement Scheme

Figure 10 Land tenure/land use in the North Mkata Plain.

By 1970–71, approximately 29,500 hectares of land in the North Mkata Plain could be mapped as sisal estates. But the figure is somewhat misleading. Prices for sisal have been fluctuating and often low through the history of the industry in Tanzania. Substantial deterioration has occurred since the early and middle 1960s. Large areas, though still planted to sisal, have been abandoned. Attempts are now being made to diversify the economy of the estates. Several hundred hectares of estate land have been, or are being, converted to crops such as sunflower, sesame, pawpaw, and sorghum. The introduction of livestock (hogs and particularly beef cattle) is also being encouraged. Plans to convert sisal estate land to rice are receiving stimulus from government allocations of monies for this purpose and from the recent construction of a rice mill in Morogoro by the National Milling Corporation and the Morogoro Cooperative Union.

It should be noted, however, that diversification is taking place for reasons additional to the economic problems of the sisal industry. In some instances it has come about as a by-product of the development and adoption of innovations in sisal cultivation. One example of such diversification is the introduction of leguminous cover crops. The object of this technique is to protect the ground between the sisal rows from the effects of sun and rain, and at the same time to increase the nitrogen content of the soils. The by-product benefits include the sale of such leguminous cash crops as soyabeans, cowpeas, and beans. As well, the legumes provide fodder for grazing cattle (Rijkebusch, 1967; Macfarlane, 1968; Lock, 1969).

Nor has diversification meant the complete abandonment of the sisal industry. Indeed, a major replantation scheme has been initiated on selected estate lands held by the government-owned Tanzania Sisal Corporation (see Illustration 9). This scheme was to have cost over seventeen million shillings (over $2.2 million) over the period of the Second Five-Year Plan (Tanzania, 1969).

Wami Irrigation Scheme
Over 7,000 hectares of land along the Wami River east of Dakawa have been set aside for a proposed irrigation scheme. As of 1971, the grant of occupancy was held by the Morogoro Region administration and no development had occurred. In 1970-71, the first land surveys were being conducted. The only natural resources inventory that had been conducted was a brief survey of soil pH data collected by the Ubungo (Dar es Salaam) staff of the Water Development and Irrigation Division of the Ministry of Agriculture.

Wami Prison Farm
The 3,500-hectare prison farm, located southwest of Dakawa, includes a valuable retraining plan in its rehabilitation program. Prisoners from this area and others are taught sound agronomic principles of crop and animal husbandry in the field. Produce from the farm is internally consumed although surpluses do at times reach outside markets. The institution is not designed, however, to play any major role in the economic development of the North Mkata Plain, although rehabilitated inmates will benefit from this 'extension' service.

Nguru Forest Reserve
This extensive reserve, of approximately 32,900 hectares, plays little part in the economy of the area. It is held by the government under the Reserved Land: Forests Ordinance (various dates) and used almost solely as a fuel reserve for the town of Morogoro. Although squatters have begun to cultivate a few hectares of the reserve, the activity is illegal and not likely to be successful for this reason and because of the inadequacy of water supplies.

National Development Corporation (NDC) Cattle Ranch (Extension)
The NDC Cattle Ranch straddles the Central Railway Line in the south central portion of the North Mkata Plain. The ranch was started in 1948 by the Tanganyika Veterinary Department with the aim of keeping and developing breeding stock for the rest of Tanganyika. It was soon found that the area was unsuitable for this purpose because of the coarseness of the natural forage and the widespread presence of the tsetse fly, so the policy of raising beef cattle was adopted. This activity was continued by the Tanganyika Agricultural Corporation when it took over the ranch in 1961, and continues to be the policy of the present administrators, the National Development Corporation, who took over in 1964.

The ranch is extensive, approximately 100,000 hectares in area, but only 43,300 hectares are found in the survey area, and much of this area is an unoccupied extension. Although impressive on a map much of the ranch is ungrazed bush and even the paddocks that are used are unimproved.

Throughout the life of the ranch, the cattle have been mostly crosses of the boron and the small East African zebu. The majority have been bred and raised on the ranch itself, although some purchases are made from the government ranch at Kongwa, 200 kilometres to the northwest. The primary market for the ranch is Tanganyika Packers, Dar es Salaam, although a few carcasses are sold to butchers in Kilosa and Morogoro.

Ujamaa Villages
By 1970-71, the North Mkata Plain had been virtually unaffected by the development of ujamaa villages. Only a few hectares had been cultivated collectively at the fledgling ujamaa village at Mvumi. There the development had been extremely slow, with the majority of the participating farmers maintaining farms (*shambas*) elsewhere. The more recent villagization policy also appears to have had little but nominal impact.

Other Largeholdings
A number of large-scale farms, some of which were originally held by government leaseholds, may be found in the North Mkata Plain. The majority of these

farms were cleared after the Second World War and came under the development conditions of the Land Regulations, 1948. Few farms failed to meet the development conditions as they have been extensively cultivated for long periods of time. The major cash crops grown are maize, millet, sunflower, sugarcane, and rice.

Of the 15,300 hectares of land in this category in the North Mkata Plain, approximately 12,000 hectares are concentrated around the intersection of the Kilosa-Turiani and Morogoro-Mvomero roads. These farms are known collectively as the Magole Farms and the East and West Wami River Farms. Rights of occupancy were granted for most of them between 1947 and 1953. Although the size of the holdings varies from 38 hectares to 400 hectares, the majority of farms are close to 200 hectares in extent.

On the whole, these are heavily capitalized farming enterprises. Though few in number, hired African field labourers play an important part in the working of the farms. The grant holders, mostly Tanzanians of Asian origin, own or share heavy agricultural machinery such as tractors and combine harvesters and market their produce through co-operatives based in Morogoro and Mvomero. In the early years of the establishment of these farms, a training program was set up to educate the grant holders and their probable hiers in modern farming techniques. Little interest was shown in the school, however. The resultant deficiency in training can be seen in the land today: plough pans that have developed through the improper use of disc ploughs; decreasing soil fertility as a result of constant cultivation in the absence of fertilizer application; and the beginnings of soil erosion on the clean-cultivated, unprotected fields.

Customary Rights of Occupancy

In general, land not held by grant in the North Mkata Plain is either unoccupied or governed by customary land tenure rules and regulations. Approximately 142,000 hectares of the plain fall into the latter class. The three major land use categories outlined below describe the different types of occupancy of the land.

Recently Cultivated Smallholdings

Historically, the North Mkata Plain has been a contact point for many tribal groups of Tanzania, and a contact point for these tribes and foreign influences: Arab slave traders, European missionaries, and European and Asian settlers. The interaction of all of these groups with the local people, not always without conflict, has significantly influenced the character and location of African systems of agriculture in the study area. Unfortunately, few of the effects of this interaction have been adequately documented, but the present pattern can be portrayed.

Prior to the present century, little of the North Mkata Plain was cultivated, either by aliens or by Africans. The three main cultivating tribes in adjacent areas, the Kaguru, the Ngulu, and the Luguru, had sought refuge from belligerent Baraguyu, Ndorobo, Masai, Hehe, and Nyamwezi in the Ukaguru, Nguru, and Luguru mountains. With colonial rule, the warfare that had been common for decades, or perhaps centuries, came to an end, allowing for settlement of the lowlands. In the North Mkata Plain this affected particularly the Kaguru and Ngulu, some of whom settled along the Kilosa-Turiani road. In fact, this settlement was encouraged by a deliberate spacing-out of the European plantations to provide for native reserves from which labour for the sisal estates could be drawn. This policy was not completely effective, however, as the Kaguru and Nguru, for the most part, preferred to maintain their autonomy and settled in the areas indicated to provide only food crops for the estate labourers, not the labour itself. As a result, most of the estate labour had to be recruited from other tribal groups in Tanganyika and from neighbouring countries.

By 1957, a large proportion of the population in the districts of Kilosa and Morogoro was non-local. In the North Mkata Plain these people were primarily estate labourers who lived on estate-held land. Tables 12 and 13 give the breakdown of these tribal groups. By 1967, the population totals in these two districts, Kilosa and Morogoro, stood at 193,810 and 316,372, respectively (Tanzania, 1969b). In the North Mkata Plain, the trend involving imported estate labour and the maintenance of local autonomy (i.e., isolated from direct labour involvement in the estate economy) by local groups is continuing.

From the 1970-71 survey, approximately 31,400 hectares of land in the North Mkata Plain can be mapped as involving smallholder agricultural activity. Most of this area is held by rights of occupancy deriving from native custom and law. The customary rights are those of the many local and non-local groups, modified to local and changing socio-economic characteristics of the North Mkata Plain. As an example of these rights, an examination of Ngulu land tenure and land use is provided in Chapter 7. At present, over twenty-five crops are grown on smallholdings in the North Mkata Plain. The primary markets are local, although some cash crops (cotton, rice, and millet) reach Morogoro markets. On the whole, however, the major portion of the produce is consumed locally.

Unoccupied or Periodically Grazed Land and the Wakwavi Settlement Scheme
Over 94,000 hectares of land in the North Mkata Plain are unoccupied or are used for semi-subsistence herding. To this may be added over 16,000 hectares of the Wakwavi Settlement used at present for periodic grazing. In total, this area is composed of the steep denuded or forested slopes of the Kaguru-Nguru escarpment, a small portion of the Kidunda Land Unit, and a major section of the

TABLE 12

Tribal population of Kilosa District, 1957

Tribe	Number	Percentage
Kaguru	50,355	32.8
Ngulu	2,007	1.3
Baraguyu	1,789	1.2
Luguru	7,196	4.7
Sagara	18,555	12.1
Other	73,396	47.9
TOTAL	153,298	100.0

SOURCE: Tanganyika (1957)

TABLE 13

Tribal population of Morogoro District, 1957

Tribe	Number	Percentage
Ngulu	25,100	9.4
Baraguyu	386	0.1
Kaguru	1,616	0.6
Luguru	175,786	65.8
Zigulu	7,777	2.9
Kwere	6,551	2.5
Other	49,819	18.7
TOTAL	267,035	100.0

SOURCE: Tanganyika (1957)

grasslands and wooded grasslands of the North Mkata Plain. It is the central grassland section of the North Mkata Plain, principally the Mkata Station-Dakawa Land Unit, that is occupied by the semi-nomadic herders, the Baraguyu (sometimes referred to as the Kwavi).

The Baraguyu people are considered to be one of the most widely dispersed peoples in East Africa, members of the group being located as far south as the Mbeya and Iringa districts and as far north as the Pare District, in Tanzania (Beidelman, 1962b). The North Mkata Plain is one of the three sites of concentration of these people, the other two being the northwestern Dodoma-southeastern Kondoa and western Pare districts. In 1957 slightly more than 2,000 Baraguyu (Kwavi) were reported to be living in the districts of Kilosa and Morogoro, the majority in the North Mkata Plain. Their present numbers in the study area would be a matter of speculation. Figure 11 locates the principal

Figure 11 Distribution of Baraguyu bomas in the North Mkata Plain.

stockades (bomas) in the area at present (see also Illustration 10). This gives an idea of distribution, but not of numbers.

The Baraguyu are pastoral people, with a diet principally of cattle flesh, milk, and blood. For the most part, they graze their herds (the numbers of which are open to even more speculation than the numbers of Baraguyu people) in the unoccupied areas along the Wami River; but conflicts often arise when they allow their cattle, donkeys, and goats to stray into the cultivated areas below the Kaguru-Kidete-Nguru escarpment. This situation usually occurs during the dry season when the central plain is unable to provide adequate forage. At present, an attempt is being made to alleviate this problem by the establishment of the Wakwavi Settlement area.

Located to the east of the Dakawa-Kwadihombo road, the 16,200-hectare Wakwavi Settlement Scheme is designed to provide water, health, and educational facilities, and therefore a permanent settlement, for the Baraguyu. Originally proposed in 1963, the scheme did not receive impetus until the allocation of Shillings 500,000 ($70,000) to it in the Second Five-Year Plan (Tanzania, 1969a). By 1971 the Water Development and Irrigation Division had completed the water distribution system (boreholes, water storage facilities, pipelines, several supply sites, and a cattle dip) and a number of Baraguyu were becoming interested in the project.

With respect to land tenure, the areas under this land use heading are considered by the Baraguyu to be communally held pastures by right of customary occupancy. These include the land slated as being part of the Wakwavi Settlement Scheme. As far as can be determined this interpretation is correct, as the legal rights of occupancy of the scheme have not been resolved. It is possible that the settlement area will come under the regulations of the Range Development and Management Act, 1964, which has been applied in other grazing areas of Tanzania, notably in parts of Masailand. This legislation was enacted to conserve and develop range areas. Among its provisions is the allowance for the establishment of ranching associations. Such an association could be formed when 60 per cent of the prospective members approved of its establishment. Under the act, all customary rights would be abolished and replaced by a grant of right of occupancy. In this way, an association, guided by a range development commission, could adequately plan and control the use of the Wakwavi Settlement land.

7

Ngulu Land Tenure and Land Use

The North Mkata Plain has been an area of intense interaction of many tribal and alien elements over the past century. Nevertheless, the members of each of the tribal groups that have occupied this area have settled in relatively discrete tribal zones. For example, all of the people interviewed in the Makuyu and Kigugu villages gave as their tribal affiliation the Ngulu. An opportunity was thus afforded to hold this parameter constant in dealing with many aspects of land-holding and land use. Unfortunately little detailed information has been recorded. The analysis below is heavily dependent on the field survey, the procedures for which have been outlined in Chapter 2, and is supplemented by reference to the few lines about the Ngulu written by Beidelman (1960, 1961, 1962a, 1962b, and 1967) in his studies of the Kaguru and Baraguyu. The discussion also draws on the data found in the short note on the Ngulu published by Ntemo (1956).

Like many Tanzanian people in the days immediately prior to the colonial periods, the Ngulu inhabited predominantly upland, mountainous areas. These sometimes harsh environments afforded protection from raiding neighbours interested in material booty and/or slaves. In the case of the Ngulu, the Nguru Mountains provided, at various times, refuge from the Ndorobo, Baraguyu, Masai, Hehe, Nyamwezi, and some of the slave-trading coastal tribes in the employ of Arab traders. Nevertheless, this point should not be overstressed. The Ngulu have also engaged in bellicose excursions while pursuing the trading activities for which they are more widely known. Their particular targets have often been their immediate neighbours, the Kaguru and Zigua.

With colonial rule and the establishment of political order, many Tanzanian tribes once isolated in upland areas were able to move to lower elevations. While some Ngulu may have followed this pattern and begun to settle lowland areas close to the Nguru Mountains soon after the establishment of German colonial rule, it would appear that general Ngulu movement of this type did not occur until much later. If one is allowed some liberty with Ntemo's (1956) distinction between present-day Ngulu and Zigua, one might suggest that: the two tribes were one at one time; the Zigua settled in lowland areas to the east of the Nguru Mountains fairly soon after German establishment of order if not before; and then the Ngulu gradually encroached on Zigua lands forcing them further eastwards. Although these are speculations, movements out of the Nguru Mountains do seem to have been of a relatively late date, and particularly when many Ngulu were forced out of the upland areas in the early 1930s as a result of famine conditions (Ntemo, 1956, and Ngulu informants).

Whatever the precise migration history of the Ngulu, Ungulu (the land of the Ngulu), while including large areas of the Nguru Mountains, today comprises a very significant portion of the North Mkata Plain and lowland areas further east and north. In more recent years, the greatest concentrations of Ngulu are found in the northwestern third of Morogoro District and the western third of Handeni District. Table 14 provides census data for the Ngulu in these districts as well as in the three other districts in which Ngulu are found.

In Table 14, Percentage Change and Average Annual Percentage Change columns are included, primarily to satisfy curiosity. Without further data (birth rates, death rates, migration information, etc.) it is doubtful whether these data deserve close scrutiny, especially when one finds that Ntemo (1956) considers that 22,625 Ngulu resided in Lushoto District in 1948, but Beidelman (1967) does not record any figure for them in this district in 1957. And vagueness is evident in Beidelman's (1967) statement that 'large numbers of Ngulu are said to be engaged as itinerant petty traders outside their homeland.'

What may be considered significant for this publication is that:

1 Ngulu, by area and by population, make up a large proportion of the people of the North Mkata Plain study area, and are therefore important for its development.
2 Ngulu have traditionally been located in upland areas and have only recently occupied the lowlands of the study area. As these people have lived in widely dissimilar environments, the land use and land tenure systems are not likely to be uniform throughout Ungulu, although there will in all probability be some similarities; and their land use and land tenure systems are likely to be in flux for this reason, if for no other.

TRADITIONAL LAND TENURE PATTERNS

In the political structure of the Ngulu, no paramount authority was recognized. Rather, small areas within Ungulu were controlled within a loose tribal structure.

TABLE 14

Ngulu population in Tanzania by districts of major occurrence, 1948 and 1957

District	1948	1957	Percentage change	Average annual percentage change
Morogoro	21,644	25,100	16.0	1.78
Handeni	17,408	20,100	15.5	1.72
Kilosa	2,853	2,007	−29.7	−3.30
Mpwapwa	1,142	1,892	65.7	7.30
Masai	n.d.	1,276	n.d.	n.d.
TOTAL	43,047	50,375	17.0	1.89

SOURCE: Compiled from Ntemo (1956), Tanganyika (1957), and Beidelman (1967)

These areas of land, whose borders were rarely fixed or defined, were thought of as belonging to various clans and lineages. As each clan was divided into many lineages, the practical implementation of land tenure rules was applied to particular areas of land associated with a particular lineage. Because shifting cultivation was the dominant form of agriculture, and because the number of members in any lineage varied, it would be impossible to give an estimate of the size of the areas so formed. No figures have been published giving such information, nor were Ngulu informants able to suggest any.

Each lineage, led by a headman (*jumbe*), settled lineage land in a more or less nucleated village within the core area of the lineage holdings. The headman, usually the founder of a lineage or one who had achieved authority by inheritance, acted as a land allocator and arbiter of land disputes. In this function, the headman was advised by the elders of the lineage, although he could apparently make major decisions against their wishes. A fee was not levied by the headman for his services as land allocator or disputes arbiter, but he could receive gifts in kind.

Each Ngulu adult male, as a member of a lineage, had the right to occupy enough lineage land to meet his subsistence needs. The amount of land required depended wholly on the size of the man's household (that is, the number of his immediate dependents). Women were also able to hold land if widowed and/or without the support of a son or a brother.

To obtain land, a lineage member could either inherit it or clear land not currently occupied. In either case, the headman oversaw the taking up of such land. This process, at times accompanied by the payment of gifts in kind, was witnessed by relatives or, in their absence, by friends or neighbours.

Traditionally, the Ngulu were a matrilineal society. Descent, and therefore succession and inheritance, have been reckoned through the female line. Lineage heads who obtained their authority by inheritance did so from a matrilineal relative. A man would receive his inheritance of land (or of material goods) from his maternal uncles rather than from a father. In the absence of a maternal uncle, inheritance would come from the closest maternal male relative, in which case a man's inheritance might be of very little value. In principle, all nephews inherited their uncle's land equally. However, in times when land was in good supply, an uncle's material goods could be substituted for land in the inheritance. Under these conditions, even if the number of nephews was large, each nephew would not necessarily expect to receive an equal share of his deceased uncle's land. The role of the headman as land allocator and arbiter of disputes took on special importance when difficulties arose over an inheritance because there was a large number of claimants and land was in short supply.

Whether land was inherited or obtained through the allocation of unoccupied land, an individual's rights in land were not exclusive, with complete powers over it. Land could be borrowed or loaned by mutual agreement with no fee being levied. But land could not be allocated to, or loaned to, a non-lineage member without the approval of the headman. Nor could land be sold to anyone, whether a lineage member or not. In the situations where settlement had become relatively permanent, however, improvements to the land required compensation upon reallocation. The continuance of rights to a particular parcel of land required no specific improvements to it, but did require continuing occupance (including fallow periods).

NGULU STUDY AREA

The Ngulu of the North Mkata Plain continue to be hoe-cultivators. 'Men burn and clear fields and do much of the planting. Both sexes cultivate and harvest. Women cook, draw water, fetch firewood, carry loads, tend children, clean houses, make pottery and baskets, and brew beer from maize, millet, sugarcane, and honey' (Beidelman, 1967). Assuming that the Ngulu of the Makuyu and Kigugu villages are characteristic, livestock does not play a part in their agricultural economy. The above statements could apply to both past and present agricultural systems of the Ngulu of the North Mkata Plain. But while there is no detailed historical information, selected land use data from the present can now be provided. Much of these data may be linked with the evolving system of land tenure.

In the surveyed areas, the Ngulu follow a system of permanent or semi-permanent land cultivation. Plots of land or fields are well marked and used by the same farmer for long periods of time. Annual cultivation over a period of ten to twenty years occurs before the land is abandoned. With no application of manure or commercial fertilizer and with a general absence of fallowing, declining yields are considered to be part of the overall agricultural cycle. Movement (shifting) occurs only after below-subsistence yields are obtained.

The majority of Makuyu farmers are not subsistence farmers in the sense that all they produce is necessarily consumed by the household producing it. Surplus crops (when a surplus occurs) and crops intended to be cash crops at the time of planting are sold directly to Africans and Asians on nearby sisal estates and on other largeholdings, taken directly to markets in Mvomero, Magole, or Turiani, or sold through the Mkundi or Mvomero agricultural co-operatives. In addition, surplus agricultural produce is bartered for Baraguyu livestock products (especially milk). In the survey of Makuyu households, it was impossible to elicit responses that would allow one to say what proportions of the principal crops cultivated (except the non-food crops) were intended for market from the time of planting. These proportions become evident only after surpluses are in fact obtained. In general, cotton, castor, and sunflower are referred to as cash crops, while maize, millet, cassava, and banana are the food crops, sold only if there are surpluses.

The annual agricultural cycle is tied closely to the seasonal distribution of rainfall. The people of the Makuyu and Kigugu villages tend to think of the wet season as being composed of two periods. Their Short Rains (*mvua za vuli*) arrive in November and extend through the end of December and into January. Their Long Rains (*mvua za masika*) arrive at the end of January or in February and end in May or June. As illustrated in Chapter 3, the statistical data on rainfall do not support a division into two periods. However, the agricultural field activities of the Ngulu do follow this pattern.

Most fields are cleared immediately prior to, or at the beginning of, the Short Rains. In the survey area, clearing generally means the burning of the residue of the previous harvest and other debris. It rarely involves the cutting of trees or bush. Seedbed preparation follows soon after in later November and early December, but may extend into January. Sowing, with priority given to food crops, is broken into two periods in an effort (possibly a vain effort, given the analysis of the climatic data in Chapter 3) to maximize the use of the available rainfall. A portion of the crops is sown in December while the remainder is sown from mid-January to early February.

Weeding is done towards the end of February and in March, while harvesting occurs from July through to September, depending on the date of planting. All of the work involved, from clearing to harvesting, is done by hand. The only tools employed are metal hoes and pangas, machetelike knives 45 to 60 centimetres in length.

TABLE 15

Makuyu and Kigugu villages: Crop proportions (per cent) by village area

Village	Millet	Maize	Cotton	Cassava	Banana	Sunflower	Rice	Castor	Other
Makuyu	41	30	19	3	1	5	0	1	0
Kipinde	38	32	6	6	6	6	0	6	0
Mkololoni	38	36	10	5	2	7	0	2	0
Muibuka	30	38	0	8	0	8	8	0	8
Milongwe	50	30	0	5	0	10	0	5	0
Mzizima	45	41	4	4	1	1	0	4	0
Chamkole	46	43	0	2	2	2	5	0	0
Kigugu	10	40	0	3	3	0	43	0	1
Madegho	13	38	3	0	0	0	46	0	0
Makuto	16	33	0	0	0	0	51	0	0
Dibamba	34	0	0	0	0	0	66	0	0
ALL VILLAGES	30	36	6	3	2	3	17	2	1

TABLE 16

Makuyu and Kigugu villages: Proportions of fields within various distance intervals from field to village where distances are measured along major footpaths

Distance intervals (metres)	Proportion of fields (percentage)
Less than 150	20.9
150–249	17.4
250–349	20.3
350–449	9.2
450–549	10.2
550–649	7.8
650–800	4.4
Greater than 800	9.8

Cropping Patterns

In the Old Central Province of Tanganyika, of which the North Mkata Plain was a part, the numbers of hectares used for the major food crops in 1945 were estimated to be: millet, 157,300; sorghum, 48,400; and maize, 7,700 (Ruthenberg, 1964). The figures for the World Agricultural Census revealed essentially the same pattern for the year 1950 (Ruthenberg, 1964). Twenty years later, the

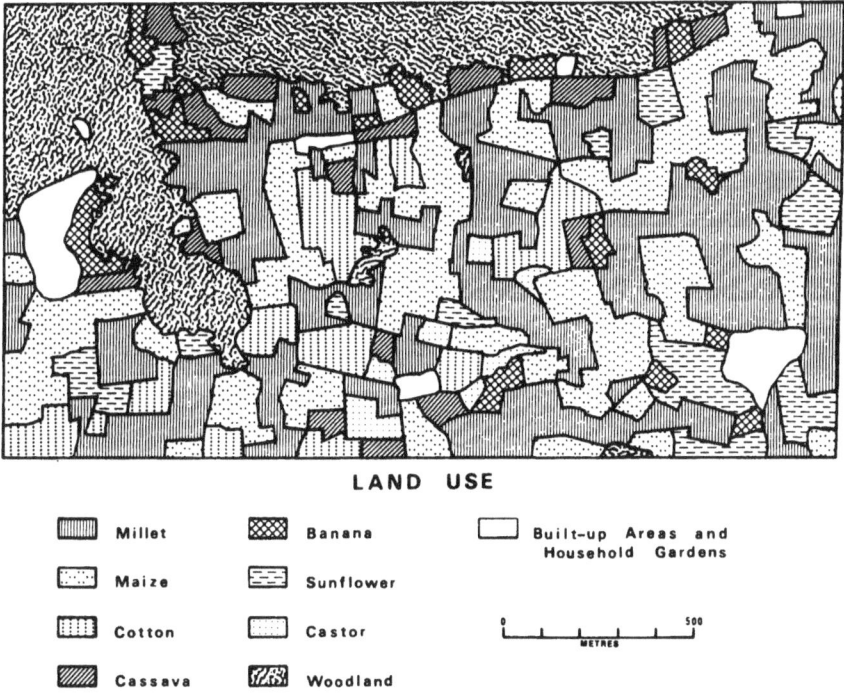

LAND USE

▦ Millet	▧ Banana	▢ Built-up Areas and Household Gardens
▢ Maize	⊟ Sunflower	
▤ Cotton	▢ Castor	0 ─────── 500 METRES
▨ Cassava	▨ Woodland	

Figure 12 Land use — selected area of the Makuyu key area.

figures derived from the Makuyu and Kigugu villages showed a very different pattern. Maize had taken a significant position in the crop distribution pattern, rivalling the position of millet, while sorghum was not grown at all. Proportional hectarage figures for maize, millet, and the other principal crops for the surveyed Ngulu villages are given in Table 15. The spatial dominance of millet and maize is well illustrated in Figure 12. This is a map of a portion of the Makuyu area showing the use of recently cultivated land.

Of special note in Table 16 are the large figures for rice in the Kigugu villages. These figures not only reflect village location along small streams and in flood-plain areas, but also the trend by government officials to encourage the planting of this crop (Tanzania, 1969a). This difference in crop proportions between the Makuyu and Kigugu villages is further reflected in such land use characteristics as farm size, field size, and the number of fields per farm. Some of these features of land use have been directly affected by land tenure changes and will be discussed in greater detail below.

Prior to a discussion of Ngulu cropping patterns, it should be observed that, in searching for a general theory of location for agricultural land use, present-day geographers have made use of the principles and methodology that were outlined by Von Thünen in the first half of the nineteenth century and re-examined by Chisholm (1962), Gregor (1970), Found (1970), Baker (1973), and numerous others. One of the essential elements of the model is the optimization of the location of land use as determined by distance. Most work along these lines can be characterized as describing the distance-decay relationships of agricultural activity around a market centre.

However, few researchers have addressed themselves to Von Thünen's postulation that these distance-decay relationships also apply within a farm itself. Exceptions to this may be found in Chisholm (1962) and in more recent articles by Found (1970) and De Lisle (1978). The consequences that Chisholm sees in this relationship, as paraphrased by Jackson (1972), are:

1 The pattern of cropping on each field on each farm should be so arranged as to maximize the net return of each field.
2 For any particular type of farming there is an optimal areal extent of holding and farm size is therefore closely related to location.
3 The distance from the central city at which a particular crop is grown is affected by the distance, within the farm, from the farmstead at which it is grown.
4 Important benefits would accrue from farm consolidation schemes.

While Chisholm (1962) provides a summary of empirical work that generally supports these consequences, Jackson (1972) questions their applicability in rural Africa. In Tanzania the evidence is conflicting. In the book edited by Ruthenberg (1968a) Jackson quotes a number of papers that counter these proposals. Hirst (1970), in contrast, has indicated that distance-decay relationships of this type do exist. Are they operative at the farm level in the North Mkata Plain? Several of these hypotheses can be examined when discussing Ngulu land tenure and land use patterns.

For example, if the Von Thünen model was operative, one would expect to observe spatial regularities in land use in the form of cultivation zones of concentric rings around the Makuyu and Kigugu villages. This hypothesis was tested by constructing a series of concentric circles around the survey villages and then enumerating the types of crops by field within these circles. The increments to the radii of the circles was an arbitrarily chosen value of 100 metres. This procedure was completed for all of the Makuyu villages but not for the Kigugu villages because the latter were not as intensively surveyed.

In examining the number of fields planted to specific crops within the concentric circles, the expected pattern does not emerge. That is, no crop is consistently dominant in any of the concentric zones from village to village. For example, maize fields make up a large portion (20 to 100 per cent) of the fields in all of the concentric zones around all of the villages, but the zones in which the number of maize fields is greater than 50 per cent are not the same from village to village. The same lack of pattern holds true for all of the other crops mapped in the survey area.

Enumerating the number of fields by crop by grouping the concentric zones also does not reveal cropping patterns in concentric rings. For example, when the distance from field to farmstead is measured by major pathways, the average distance is 350 metres. This figure, rounded off to 400 metres, could be used to define those concentric zones that were *near* (within the first four zones from the village) and *far* (beyond the first four zones from the village). Again no specific crop was found to dominate either the near or the far zone.

Given that cash-cropping is not a major activity of the Ngulu, it might be expected that where specific crops do not make up concentric cultivation zones, this type of zoning might emerge nevertheless if the crops were grouped as either cash crops or food crops. Once again, this is not the case. For example, of the 529 fields surveyed in the Makuyu villages area, 84 per cent were planted to food crops. These food crops were equally distributed in both the near and far zones. The cash crop fields had a similar distribution. Overall, the pattern is as follows: food crops – near zones, 41 per cent of all fields, and far zones, 43 per cent; cash crops – near zones, 7 per cent and far zones, 9 per cent.

When the area of specific crops or groups of crops rather than the number of fields is used in this analysis, the pattern is the same, that is, no concentric cultivation zones emerge. Therefore, for the Ngulu villages surveyed, the data do not support the Von Thünen hypothesis in terms of crop type. Similar conclusions would likely be arrived at if field output were measured by yield or value, but insufficient data of this character were collected to follow through with such analyses.

The Von Thünen model was based on the principle that the economic rent of a field declined with distance from farmstead. But, as the discussion below indicates, distances from farmstead to field in the survey area are small. Therefore distance minimally influences economic rent and is one of the main reasons why the cropping patterns of the Ngulu area do not conform to the concentric zones of the Von Thünen model.

In the Ngulu survey area the average distance from farmstead to field, where distance is measured by major footpath, is approximately 350 metres. But this is a measure which is extremely variable as illustrated in Table 16. In fact, the median distance is less than 350 metres. Given this average figure, however, an estimate can be made as to what proportion of production time is taken up in travelling to and from the fields.

All travel to and from fields in the survey area is by foot. In estimating distances from house to field in Tanzania, Hirst (1970) multiplied the number of hours it took a person to walk this distance by the factor 2.5. This procedure gave an estimate of the distance measured in miles or a walking rate of 4,000 metres per hour. At this rate, the return journey from field to village would require approximately ten minutes if the distance to be covered was 350 metres. A farmer's labour time actually spent in the field would be reduced by ten minutes each day if the return journey were made only once. In an eight-hour working day, for example, 2 per cent of the production time would then be consumed by walking to and from the fields. If the working day were five hours, as suggested by Ruthenberg (1968b), the proportion would increase to slightly more than 3 per cent. Figures of 2 and 3 per cent represent relatively insignificant proportions of a person's working day and therefore would have little influence on the economic rent of land.

Farm and Field Size
The size of holdings, whether by field or by total holding, of the farmers of Makuyu and Kigugu is extremely variable. Fields range from 0.10 hectare to 2.83 hectares as measured by the technique outlined in Chapter 2. These figures do not include the smaller garden plots located just outside the farmsteads. Combinations of fields of this size provide total holdings ranging from 0.30 to 6.88 hectares in size. A more complete outline of these variations by village is provided in Table 17.

The average size of a holding in the Ngulu survey area is 1.73 hectares, although considerable variation exists as illustrated in Table 17 and Figure 13. This size of holding is common to many areas in Tanzania where the only farm implements are hoe and panga (Ruthenberg, 1968b). Given the level of technology, these holdings cannot be considered small. To illustrate this point, the amount of labour required for the cultivation of these holdings is examined.

Labour requirements (measured in man-days per hectare) of the main agricultural work activities in the Ngulu survey area were estimated to be: clearing and seedbed preparation, 51; planting, 15; weeding, 29; and harvesting, 39. On the average, a total of 134 man-days per hectare are required. These values, although rough estimates derived from the questionnaires administered to the farmers of the Ngulu survey area, are comparable to those in other areas of Tanzania. For example, Attems (1968) reports figures of 49, 12, 24, and 37 man-days per hectare for the respective activities of clearing and seedbed

TABLE 17

Makuyu and Kigugu villages: Farm and field size characteristics

Village	Farm size (hectares)			Field size (hectares)		
	Mean	Standard deviation	Range	Mean	Standard deviation	Range
Makuyu	2.07	0.97	0.40–5.46	0.67	0.43.	0.20–2.02
Kipinde	1.73	0.42	0.80–2.63	0.68	0.73	0.10–1.21
Mkololoni	1.80	0.75	0.40–2.63	0.64	0.32	0.10–1.62
Muibuka	1.49	0.53	0.61–3.24	0.67	0.27	0.10–1.62
Milongwe	2.54	1.65	1.62–6.88	1.20	0.74	0.10–2.83
Mzizima	1.84	0.66	0.30–2.63	0.90	0.36	0.10–1.21
Chamkole	1.77	0.76	0.81–4.05	1.23	0.37	0.10–2.02
Kigugu	1.10	0.64	0.40–4.05	0.81	0.26	0.10–1.62
Madegho	1.09	0.64	0.20–2.42	0.79	0.25	0.20–1.21
Makuto	0.76	0.30	0.40–1.21	0.79	0.07	0.40–0.61
Dibamba	1.01	0.61	0.40–1.62	0.61	0.17	0.40–0.81
All villages in Makuyu area	1.89	0.82	0.30–6.88	0.86	0.46	0.10–2.83
All villages in Kigugu area	0.99	0.55	0.20–4.05	0.68	0.19	0.10–1.62
ALL VILLAGES	1.73	1.38	0.30–6.88	0.78	0.38	0.10–2.83

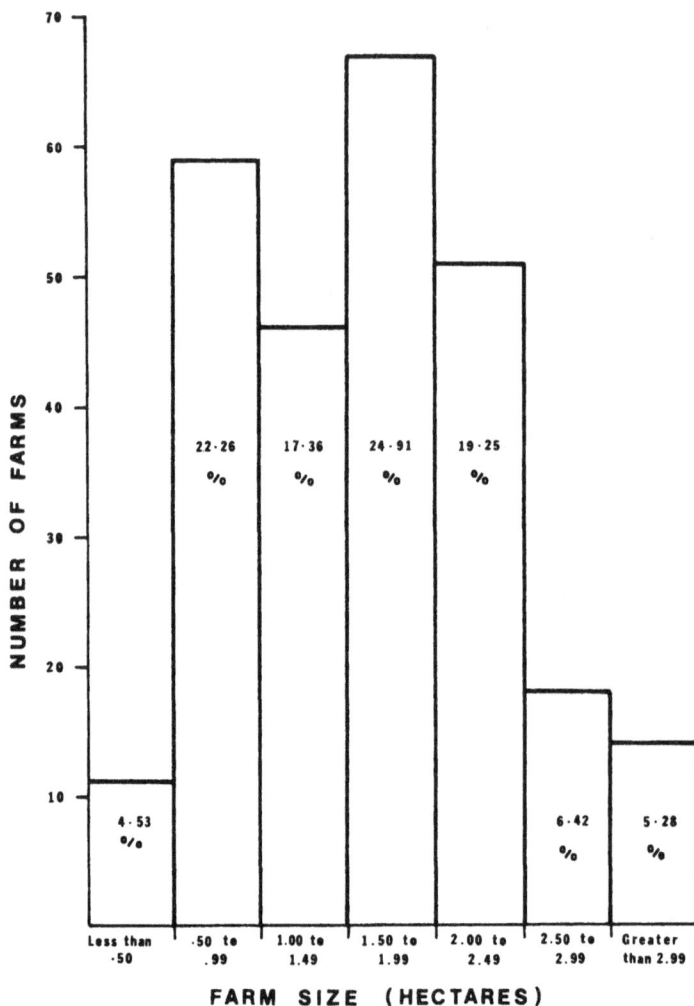

Figure 13 Frequency distribution of farm sizes in the Ngulu survey area.

preparation, planting, weeding, and harvesting. McKay et al. (1970) give similar data. Their estimates were, in terms of man-days per hectare: clearing and seedbed preparation, 64; planting, 8; and weeding, 39.

In the latter study, McKay et al. (1970) found that labour requirements of this order of magnitude were such that a household having 2.2 man-equivalents (ME, see below for definition) available for these farm activities would not be capable of cultivating much more than 2 hectares. That is, the clearing of more land than this at the time when this activity can be done casually (that is, in the dry season) would be wasted because the other activities conducted at times when a degree of speed is required (in order to maximize the available moisture) could not be completed.

The concept of *man-equivalents* has been widely used in Tanzania as a means to measure the available household labour. Introduced by Collinson (1963), the measure aggregates household labour availability by evaluating the work efficiency of different age and sex groups. The base value of 1.00 ME would be provided by a male between the ages of 19 and 50 years inclusive. By comparison, a male aged 15 to 18 years or over 50 would have a labour efficiency rating of 0.67 ME. For women of comparable ages the ratings would be 0.67 and 0.50, respectively. Children aged 10 to 14 years, male or female, would be given a rating of 0.25. Thus, if a household were composed of a 35-year-old male, a 32-year-old female, and an 11-year-old child, it could provide 1.92 ME of labour.

In the survey of the Makuyu and Kigugu villages, household occupants were enumerated either as adults (15 years of age and older) or as children. The aggregation technique outlined above cannot be precisely applied. However, an average man-equivalent figure of 1.92 per household would probably not be far off the mark (if not slightly high), considering the fact that in the Ngulu survey area the average household is composed of 1.98 adults and 1.78 children.

Since the ratios between available labour and holding size (1.92:1.73 and 2.20:2.02, respectively, or 1.1 ME per hectare) in the present study area are much the same as in that of McKay et al. (1970), the Ngulu are probably cultivating as much land as their labour availability allows. They would certainly be doing so if, as Ruthenberg (1968b) claims, the man-days per hectare requirement is incorrect. In the calculation of these figures in Tanzania, it is generally considered that the norm in working hours is between 1,800 and 2,000 per year, derived from a man-day of 8 hours. Ruthenberg (1968b) claims that this is a fictitious norm because of the low efficiency per hour of work and should be translated into a 5-hour man-day. If the situation is similar in the Makuyu and Kigugu villages area, then a total of approximately 214, not 134, man-days per hectare would be required. With an available labour component of 1.92 ME, the Ngulu would find it extremely difficult to cultivate more than the average 1.73 hectares of land.

To further illustrate the relationship between farm and field size and the number of household occupants, a number of linear correlation coefficients were calculated. Correlation coefficients between other land use characteristics of the Ngulu survey area were also determined and these are presented as well in Table 18.

As one would expect, the size of holdings (X_2) and the size of fields (X_3) are interrelated and when considered together yield a correlation coefficient of 0.85. These same areas of land are also related to the number of persons in a household (X_5). When the size of holdings (X_2) is regressed with family size (X_5), it yields a correlation coefficient of 0.53; when the size of fields (X_3) is regressed with family size (X_5), a similar value for the linear coefficient is obtained, 0.52. A relationship does exist therefore between size of holding and family size. But as these data indicate that family size can be used to explain only about 30 per cent of the variation in size of holdings, it is not comparable to the correlation coefficients of over 0.90 obtained in Uganda (Jackson, 1972).

This figure of 0.90 is used by Jackson (1972) with other data to demonstrate that in Africa the size of holding is not related closely to location with respect to farmstead, a consequence which should occur given the Von Thünen hypothesis. The data from the present study do not altogether support Jackson's views — nor, with correlation coefficients only slightly above 0.50, do they refute them. But an analysis of the relationships between distance and field size does add some weight to his arguments.

For example, given the average size of fields per village in Table 17, the number of fields that could be expected to be found within the 100-metre concentric zones around a village can be determined. These expected figures can then be compared with the actual number of fields observed in these zones. Using a chi-square evaluation, it was found that the differences in expected and observed numbers of fields per concentric zone are significant at the confidence limits of 90 per cent, 95 per cent, and 99 per cent.

This conclusion is, of course, intuitively obvious given the ranges of field sizes. No significant difference would occur except where all of the fields were the same size. But, does the difference derive from the fact that field size increases or decreases with distance from the farmstead? To examine this relationship, linear correlation coefficients were again employed, but only for the Makuyu villages.

Distance was measured in two ways: from the centre of a field to the centre of the farmer's village by straight-line distance, and from the centre of a field to the village by major footpath. For the Makuyu villages, where size (in hectares) was considered a function of straight-line distance (in metres), linear coefficients range from -0.12 to 0.21. With distance measured along footpaths, the range of coefficients is from -0.37 to 0.32. Distance is not, therefore, one of the variables that plays a major role in determining field or farm size. And, in support of Jackson's arguments, it is less significant than family size.

One of the more interesting relationships drawn out in Table 18 is that between holding size and crop type. Throughout the survey area, farm size tends

TABLE 18

Makuyu and Kigugu villages: Matrix of linear correlation coefficients comparing selected farm characteristics*

	X_1	X_2	X_3	X_4	X_5	X_6	X_7	X_8
X_1	1.00							
X_2	0.61	1.00						
X_3	0.18	0.85	1.00					
X_4	0.79	0.17	-0.28	1.00				
X_5	0.76	0.53	0.52	0.56	1.00			
X_6	0.57	0.84	0.72	0.25	0.67	1.00		
X_7	-0.18	0.20	0.24	-0.21	0.35	-0.91	1.00	
X_8	-0.62	-0.87	-0.63	-0.35	-0.91	0.73	0.18	1.00

* Where X_1 is the mean number of fields per farm per village, X_2 the mean area of farms per village, X_3 the mean area of fields per village, X_4 the mean number of adults per farm per village, X_5 the mean number of persons per farm per village, X_6 the percentage of village area in millet, X_7 the percentage of village area in maize, X_8 the percentage of village area in rice.

to increase with an increasing proportion of land cultivated in millet; and tends to decrease with an increasing proportion of land cultivated in rice. The same relationship holds true for the variation in field size (see Illustrations 11 and 12).

Although a detailed analysis of this relationship was not undertaken, several explanations may be offered to account for it. In the first instance, it is highly likely that the labour requirements per hectare for rice are greater than those for millet. That is, households with the same ME could cultivate a larger millet area than rice area. Coincident with this explanation is the observation that family size in the Kigugu area is often slightly below the average for the entire Ngulu survey area. This fact too would lead to smaller holdings and field sizes.

An additional explanation that needs further examination is the relationship between yield and food value. In his generalized account of carrying capacity in Tanzania, Moore (1971) used the following figures for the districts of Kilosa and Morogoro: *Yield:* millet, 1200 kg/ha; rice, 1500 kg/ha. *Food value after preparation:* millet, 320 calories/100 g; rice, 354 calories/100 g. If these figures are correct, it would require approximately 1.3 hectares in millet to produce the same food value as 1.0 hectare in rice.

Land Acquisition and Fragmentation

The most significant changes in land tenure practices from the traditional ones described earlier in this chapter have to do with the process involved in acquiring land and the types of land acquisition.

With respect to the process of acquisition, the headman of a village, sometimes in consultation with other elders, still has a leading role in the allocation of land. However, in some villages, especially those in the Makuyu area, the headman's authority in this respect has been challenged by the establishment of village development committees. Village development committees were created in 1963 in order to develop and organize self-help schemes at the village level. Although village headmen were often members of these committees, they were outnumbered by younger members desiring to push forward schemes of their own which often required the control of land acquisition.

In the survey of the Ngulu villages, the role of the village development committees as land allocators was not intensively examined. In conversations with the farmers and headmen of the area, the proportion of fields allocated under the influence of the village development committees was simply noted. The proportion is small (less than 10 per cent), but the power of the committees seems to be increasing. This statement is based only on the detection of some unhappiness on the headmen's part because of the loss of some of their powers.

It should be noted with respect to the village development committees that they are now closely linked with the government party. In the survey area, the party's *cells* were introduced in 1965 and 1966. Like the other party cells organized throughout the country, a cell grouped ten households together. These households elected a leader to act as the main agent of communication between the cell, the government party, and civil service structures. The leaders of the ten-house cells now make up the membership of the village development committees.

In the Ngulu survey area there is a third influence in the process of land allocation. This is the local agricultural extension officer (*bwana shamba*) who oversees much of this process. The representative is present at the time when the headman shows a petitioner a piece of land. Since he is there nominally to provide agronomic advice, the *bwana shamba* may contradict the rulings of the headman.

With two additional parties influencing the process of land allocation, the headman's traditional authority is gradually being eroded. But no evidence of sharp conflict between these powers was gathered, either because it does not exist or because the study did not set out to specifically gather such data. The potential for conflict is there, however, and may require some attention in the future.

Table 19 indicates that the traditional types of land acquisition, inheritance and clearing, remain the dominant types within the Ngulu villages area. In some villages, however, a significant proportion of the land has been acquired by borrowing and renting. Where these are the means of acquisition, no authority

TABLE 19

Makuyu and Kigugu villages: Proportion (per cent) of fields by type of acquisition

Village	Inherited	Cleared	Borrowed	Rented	Other
Makuyu	100	0	0	0	0
Kipinde	100	0	0	0	0
Mkololoni	85	15	0	0	0
Muibuka	20	40	40	0	0
Milongwe	78	22	0	0	0
Mzizima	40	40	17	3	0
Chamkole	50	50	0	0	0
Kigugu	22	46	16	0	16
Madegho	12	44	8	0	36
Makuto	60	40	0	0	0
Dibamba	50	50	0	0	0
ALL VILLAGES	71	19	5	1	4

oversees the transaction which is completed between the two or more farmers involved. And, despite the ban on such dealings in Tanzania, a number of parcels of land have been acquired by purchase (see category Other in Table 19). As the purchase of rural land is illegal, the farmers who owned land were unwilling to discuss any details of their transactions. The combination of these types of land acquisition has led to a degree of holding fragmentation in all of the surveyed village areas.

Ngulu inheritance laws are primarily responsible for the land fragmentation that is going on in these areas. But compounding some of the difficulties that arise from the traditional land inheritance practices is the fact that the Ngulu of the survey area are more and more taking on some of the characteristics common to patrilineal societies. That is, sons are beginning to inherit land from their fathers, and in some cases from their paternal uncles. At the same time, they do not give up entirely their matrilineal claims. As the situation is not clear-cut, the acquisition of inherited land often needs the assistance of an arbitrator. In the Makuyu villages area, the headman tends to support matrilineal claims, while patrilineal claims seem to be more persuasive if the arbitrator is either the village development committee or the *bwana shamba*. In either event, if unchecked, fragmentation will probably increase to a degree greater than that illustrated in Table 20.

If one were to look at the land use map of the North Mkata Plain (Figure 10), one might ask why fragmentation occurs at all, given that large areas are unoccupied. Two explanations are offered with respect to this question.

TABLE 20

Makuyu and Kigugu villages: Numbers of non-contiguous fields per holding

Village	Mean	Standard deviation	Range
Makuyu	3.10	1.46	1–7
Kipinde	2.52	1.13	1–5
Mkololoni	2.81	1.30	1–5
Muibuka	2.22	1.13	1–5
Milongwe	2.11	1.45	1–5
Mzizima	2.04	0.96	1–4
Chamkole	1.44	0.50	1–2
Kigugu	1.37	0.59	1–3
Madegho	1.39	0.59	1–3
Makuto	1.25	0.43	1–2
Dibamba	2.00	1.00	1–3
All villages in Makuyu area	2.32	1.13	1–7
All villages in Kigugu area	1.50	0.65	1–3
ALL VILLAGES	2.22	1.28	1–7

First, the environments that are cultivated by the Ngulu are dominated by the soils of the Makuyu, Magole, and Kimamba series. In general, these soils are relatively easy to cultivate by hand implements — much easier than the often flooded, fine-textured soils of the central portion of the study area, in the Mkata Station-Dakawa land unit. As well, the cultivated portions of the North Mkata Plain are coincident with those portions receiving relatively more, and more reliable, rainfall. There is, therefore, a relative, if not absolute, shortage of land suitable for cultivation. With a growth in population and the encouragement of government to expand cash crop acreage, this relative land shortage will likely become more acute without technological improvements.

Secondly, where land is suitable for cultivation, it may not be available. That is, the land may be held under customary rights of occupancy by a clan or lineage not suffering land shortage. This situation only aggravates the relative land shortage difficulties of the area. Even if the suitable land is expropriated by the government, the people may not wish to move there because of their un-familiarity with it and the possibility that it will be located some distance away from their relatives. The only apparent alternative, therefore, is for the Ngulu to continue their traditional practices which lead to greater holding fragmentation.

The differences in degree of fragmentation between villages are in large measure related to the ages of the villages. In general, where the villages are older, and the rules of inheritance have been operative over a longer period of time, the number of non-contiguous fields increases. Although no specific data for the establishment of the villages could be determined, Makuyu, Kipinde, Mkololoni, and Milongwe are longer settled than the other villages in the Makuyu area. As indicated in Table 19, most of the fields in these village areas were acquired by inheritance (ranging from 78 to 100 per cent, compared to the range of 20 to 50 per cent in the other Makuyu villages). In these older villages the mean number of fields per holding is greater than that found in the more recently established villages. A difference of means test indicated that the differences were generally significant at a confidence limit of 95 per cent.

This pattern of an increasing number of fields per holding with village age may also apply in the Kigugu area, but the sample there was too small to obtain differences that were significant. For example, of the Kigugu villages, Makuto and Dibamba are older than either Kigugu or Madegho. But the number of fields per holding in the younger villages is slightly greater than the number found for Makuto, although not for Dibamba. When the Kigugu villages are grouped with more recent Makuyu villages, however, the general relationship emerges once again. The Kigugu villages are approximately the same age as the more recent Makuyu villages, if not younger.

In areas where land is in short supply, fragmentation implies not only an increase in the number of parcels of land per holding but a decrease in the size of fields. Generally, in those village areas that have been long-established, the number of fields is greater and the size of fields smaller than in villages established more recently. This relationship holds especially where inheritance rules have been applied over a long period of time and where the land inherited borders streams. In close proximity to streams, land is better watered and more desirable, and rights to it are jealously guarded.

The fragmentation of holdings is frequently portrayed as a problem that leads to agricultural inefficiencies (Chisholm, 1962). That is, mechanization, irrigation, the efficient application of fertilizers and pesticides, and other technological improvements are difficult to introduce if holdings are excessively fragmented. And labour inefficiencies increase with increasing fragmentation when a large proportion of a labourer's time is spent travelling to and from the fields. As an example, Chisholm (1962) cites a study in which the author concluded that when the time required to cart manure to fields was in excess of one hour, 'the amount of manure used declined sharply.' Consequently, yields tend to decline when distances to fields increase.

In the Ngulu survey area, it is open to question whether the adoption of improved technology would be inhibited by fragmentation. Fragmentation does exist, but the land is relatively flat and fields are not marked with permanent objects except, on occasion, by trees at the corners. If the farmers grouped themselves into co-operatives, like the village programs that are supported by the government, the use of tractors, and certainly of oxen, would be possible. And this kind of technology would be quite compatible with present field patterns if the inhabitants wanted these to be maintained. However, other forms of technological improvement, such as irrigation development, would probably require a major reshuffling of fields and consolidation.

A degree of labour efficiency might also be gained through consolidation. In most cases the distance from farmstead to field by footpath is several times as great as the straight-line distance because of the meandering of the footpaths. Consolidation would allow for the straightening out of the access paths.

Consolidation would require major changes in the land use patterns of the area and in the land tenure practices of the Ngulu. When asked whether they would favour consolidation, most Ngulu responded with disapproval. Consolidation would be a difficult task here without a major campaign in support of it and a major reorganization of the land tenure system.

But is consolidation a desirable goal? In some areas of Africa, while results do not give a straight yes or no to the question, consolidation has been criticized. In evaluating the consolidation programs in the Kikuyu-held areas of Kenya, Sorrenson (1967) concluded that consolidation 'failed to pave the way for improving the lot of all but a small minority of the landed class.' On these same programs in Kenya, Sorrenson (1967) states that they have 'emphatically been a political rather than an economic venture.' Elsewhere, speaking to the question of labour inefficiency arising out of time needed for travel to and from fields, Brock (1969) entertains the notion that this time 'may have positive advantages as an escape from relatives, social contact and information en route, etc.'

There is also the ecological argument against consolidation. One line in this argument views fragmentation as desirable because, if the fields are distributed among different local environments, at least one field is likely to be productive. That is, if conditions in one location in a particular year turn out to be poor, the farmer has a chance to recoup his losses at another location. This argument would have little weight in the survey area as the fields of an individual farmer are generally distributed throughout an area that varies little ecologically.

Another line in the ecological argument is that taken by, for example, Igbozurike (1971). He holds the view that ecosystems gain stability through complexity. Consolidation would tend to increase the size of fields planted in one crop. Ecological stability would be lost through the adoption of monoculture practices. Conversely, intercropping, that is, a number of crop varieties per field or a number of crops in small fields with contiguous fields being in different crop types, as in the Ngulu survey area, is ecologically more efficient. As

Igbozurike (1971) points out, 'the juxtaposition of so many different kinds of plants tends to minimize the incidence of pests and diseases, many of which are virulent when acre upon acre is given over to one crop.' The analysis of yield data from the tropics by Webster and Wilson (1966) supports these arguments. They conclude that 'the limited experimental evidence available suggests that mixed cropping has advantages with many crops.'

It is obvious that there are many arguments – social, economic, political, agronomic – for and against consolidation. If consolidation of the fragmented holdings of the Ngulu survey area and others in the North Mkata Plain is attempted, considerable attention, beyond the scope of this book, must be given in weighing the pros and cons of these arguments.

PART IV

Summary and Conclusions

8

Agricultural Development of the North Mkata Plain: Problems and Prospects

Land use planners working in Tanzania are well aware of the need for an integrated approach to agricultural development planning (see, for example, Helleiner, 1968). They are aware that a plan that is economically feasible, on whatever scale, may still fail if relative social and ecological factors are not adequately considered. Berry (1968), for example, lists the following items that must be taken into account for land use planning to be successful: 'current uses of land, so that it [the land] may be workable in terms of the present socio-economic conditions of the region concerned; land potential and the availability of water resources, so that potential for agriculture may be weighted against the demands for other possible use; existing or planned development of infra-structure, otherwise effort may prove to be ineffective through lack of communication and/or lack of markets.'

Unfortunately, development and development planning in most parts of Tanzania are hampered by the lack of basic data on three types of factors — social, ecological, and economic. For example, Berry's call in 1968 for land use and land potential maps at a scale of 1:250,000, although repeated many times since, has been answered in only a few scattered areas in Tanzania (Berry, 1968, 1971a; and Berry et al., 1969, 1970).

One of the primary aims of the research embodied in this paper was to gather, and then evaluate, some of the basic data required for the agricultural development planning of the North Mkata Plain. It should be noted, however, that in examining land capability (an ecological factor) and land tenure (primarily a social factor), economic factors were not emphasized. It should also be noted that land capability and land tenure are but two of the ecological and social factors that could be examined for the process of agricultural planning. But they are considered to be extremely important elements of Tanzanian agricultural systems, and the final observations and recommendations that follow should at least provide a solid base for further research in the North Mkata Plain and elsewhere in Tanzania.

LAND CAPABILITY

As illustrated by the text and the land capability map (Figure 9), the majority of the soils of the North Mkata Plain have physical characteristics that limit their potential for cultivation. Only 23 per cent of the survey area has been mapped as belonging to capability division A. It must also be remembered that the soil series of this capability division will include within their mapped boundaries some soils that would not be rated this highly. The converse of this distribution (that is, that physically capable soils are found within areas mapped as other than division A) occurs, of course, but the extent of these better soils is relatively insignificant. In addition, supplementary analytical soils data have demonstrated that the fertility of most of the soils of the survey area is generally low.

Thus, the physical and chemical soil characteristics of more than three-quarters of the North Mkata Plain must be considered to be obstacles to future agricultural development. But the degree to which land capability will limit development varies spatially and in some instances may be modified by other ecological factors, particularly rainfall. The problems and prospects associated with these ecological considerations are summarized below for each land unit.

Agriculture is severely limited by the physical capability of the land in the highland land units — *Nguru*, *Ukaguru*, *Kidete*, and *Nguru ya Ndege*. Throughout these areas soils are shallow, often severely eroded, and slopes are very steep. Agriculture, whether for the production of crops or livestock, should be forbidden. At the time of the field survey, the Nguru ya Ndege land unit was included in a reserved area, set aside for woodland that could be used to supply fuel. This reserve should be maintained, and it is recommended that the other highland land units be similarly protected. Soil erosion in these areas could be reduced by prohibiting agriculture and by controlling the felling of the woodlands. The maintenance of a cover of vegetation in these areas would also increase the water-holding capacity of the land, thereby reducing the severity of flooding in the lowlands.

In the *Kidunda Land Unit*, the most limiting soil features are shallowness and poor drainage. But this area includes a large proportion of the Vilanza soil association which, on the basis of physical soil characteristics, has been evaluated as capability division A land. On ecological grounds, agricultural use of these soils is limited owing to the lack of dependable supplies of water, from either rainfall or surface sources. Cultivation is also restricted because of the fact that much of this mapping unit is within the Nguru Forest Reserve, which extends north and west from the Nguru ya Ndege land unit. Pastoral activities in the Kidunda land unit are restricted not only for the same reasons as cultivation but also because of the presence of tsetse fly. Taken together, these ecological characteristics must be viewed as very serious obstacles to agriculture, with little prospect for development of the land in the near future.

Elsewhere in the world, soils similar to the vertisols of the *Mkata Station-Dakawa Land Unit* are highly regarded for agriculture (Dudal, 1965). But for their management they usually require mechanized implements (which are in relatively short supply in the survey area) because of their fine texture and firm consistence. For the same reason they often require artificial drainage – an expensive proposition. In addition, the vertisols and associated soils in the North Mkata Plain are generally found in areas having long periods of drought alternating with prolonged periods of inundation. In addition to mechanization and artificial drainage, therefore, the development of these soils for arable use would entail relatively large-scale irrigation and flood control works. Such improvements would require funds of a magnitude not likely available from either the local populace or the Tanzanian government.

In the near future, the soils of the Mkata Station-Dakawa land unit will continue to be used for livestock grazing. To protect and preserve the grasslands in this area, it would be desirable if existing legislation were employed to establish range development commissions. These commissions (or perhaps one), in conjunction with the National Development Corporation Cattle Ranch, could then set limits to the number, and regulate the distribution, of cattle so as to prevent overgrazing. They might also be able to improve the quality of the pastures by implementing planting programs similar to those that have proved successful in other areas of Tanzania (Uriyo, 1966). Certainly the expansion of quality pastoral activities seems to be highly desirable in a country where livestock products provide an extremely small proportion of the peoples' diet.

The overall capability of the land for agriculture in the *Kilosa-Turiani Land Unit* is greater than that in the other land units. This greater capability has been indicated in the text and on the land capability map, especially with reference to the Magole and Makuyu mapping units. It can be seen more clearly by visually comparing the land capability map with that of the land use of the survey area (Figure 10), in conjunction with the data provided in Tables 21 and 22. This is the land unit in which most of the cultivation in the North Mkata Plain is carried out. It is not unreasonable to suggest that this concentration of arable activity has come about because of favourable ecological characteristics.

Relative to other portions of the survey area, the ecological characteristics of the Kilosa-Turiani land unit that have allowed for the evolution of this land use

TABLE 21

Proportion (per cent) of capability division land occupied by various land use categories in the North Mkata Plain

Land use	Capability division		
	A	B	C
Sisal estates	11	11	8
Wami Irrigation Scheme	4	14	0
Wami Prison Farm	0	16	0
Nguru Forest Reserve	18	8	16
NDC Cattle Ranch	1	16	0
Other largeholdings	38	2	0
Recently cultivated smallholdings	20	7	20
Unoccupied or periodically grazed	6	11	56
Wakwavi Settlement Scheme	2	15	0
TOTAL	100	100	100

TABLE 22

Proportion (per cent) of land use categories occupying land in various capability divisions in the North Mkata Plain

Land use	Capability division			
	A	B	C	TOTAL
Sisal estates	25	71	4	100
Wami Irrigation Scheme	10	90	0	100
Wami Prison Farm	0	100	0	100
Nguru Forest Reserve	42	50	8	100
NDC Cattle Ranch	2	98	0	100
Other largeholdings	87	13	0	100
Recently cultivated smallholdings	44	46	10	100
Unoccupied or periodically grazed	13	60	27	100
Wakwavi Settlement Scheme	5	95	0	100

pattern are soils having fewer physical limitations to cultivation and a greater and more dependable supply of rainfall. This land unit is also relatively free of tsetse fly, which can be a serious health hazard to humans as well as to livestock. These same ecological factors suggest that the Kilosa-Turiani land unit could be expected to respond quickly to development inputs.

Almost the whole land unit is already being used for cultivation. And, as there is so little capability division A land that is unoccupied here, the possibility of agricultural development by the expansion of the area under cultivation is extremely limited. Development, then, will come about only through changes in land use patterns and management practices.

A number of ecologically feasible changes that could be implemented are: the adoption of high-yielding crop varieties, the use of fertilizers, an increase in the density of plants per field, and the selection of planting dates that would maximize the utilization of natural moisture supplies.

Changes of this sort would require stepped-up campaigns by agriculture extension officers (many of whom are inadequately trained) to convince the local farmers of the benefits that could be derived. The soils of the area are capable, generally, of responding to these inputs. It would be useful, though, if there were experimental fields scattered throughout the area in order to confirm these assumptions. The test fields would also be valuable for demonstration purposes. At present, most farmers in the area have to travel a long way by foot to observe the developments at the Central Research Station at Ilonga. Adoption of new techniques would likely occur sooner if test fields were more accessible. Particularly valuable for African cultivators would be experimental plots located near Kwadihombo, between Mvomera and Magole, between Kidete and Msowero, and between Msowero and Mvumi.

Irrigation is an ecologically feasible agricultural practice in the Kilosa-Turiani land unit. Most of the soils in this area drain freely or could be artificially drained with little difficulty. Topographic characteristics are also favourable as slopes are commonly less than 5 per cent and the land is not dissected by many deep stream channels or gullies. Many of the tributaries that arise in the small valleys of the Ukaguru and Nguru mountains could be dammed to form reservoirs from which water for irrigation could be drawn. Such small-scale supplementary irrigation would decrease the hazard of drought both at the beginning and at the end of the growing season. In addition, the reservoirs could be used as a source of drinking water for both humans and livestock; they would also serve to prevent, or at least minimize, flooding.

But while irrigation may be ecologically feasible, its implementation would require advanced, detailed hydrological studies in order to ascertain exactly what amounts of water can be dependably supplied. Naturally, economic studies would also be required. The success of even small-scale irrigation schemes would also be highly dependent on the local farmers' being trained adequately in the techniques of what is a most exacting form of agriculture.

LAND TENURE

Most of the land use improvements recommended or noted above for the Kilosa-Turiani land unit could be adopted in the North Mkata Plain, if ecologically feasible, without being inhibited by the land tenure systems. Improvements such as the application of fertilizers, specifications for date of planting, or denser crop spacing do not require land tenure changes. Even mechanization need not be inhibited by customary tenure practices if accompanied by the co-operative efforts of smallholders.

But a major change such as the introduction of irrigation agriculture would probably require significant revision of customary tenure regulations in order to be successful. In the Ngulu study area, no direct link between soils or land capability and land tenure characteristics were discovered. Rather, tenure rights are most jealously guarded where land is in close proximity to stream channels. The key here seems to be water rather than soil quality. In this, and no doubt in other customary tenure areas, the holdings on these lands are becoming fragmented at a faster rate than holdings that are not near streams. This fragmentation is occurring because of the desirability of these better-watered lands plus the fact that patrilineal rules of inheritance are being juxtaposed on matrilineal rules. In this situation one can imagine that the fragmentation of holdings would increase along irrigation channels as well.

One way to prevent the potential excessive fragmentation postulated above would be to consolidate holdings and then allow acquisition (whether by inheritance or by other means) of holdings, not of individual fields. Another procedure, in line with some of the villagization experiments in Tanzania, would involve the abolishment of holdings per se, the land being cultivated in a wholly communal manner. Either form of consolidation would facilitate the technical aspects of the introduction of improvements such as irrigation, mechanization or the use of oxen, or the application of fertilizers and pesticides. But either form of consolidation would probably be met with resistance from an African populace which, while not recognizing individual ownership, is often closely attached to particular parcels of land. The benefits of consolidation would also have to be weighed carefully against the ecological and social arguments against consolidation that have been indicated in Chapter 7.

More immediate challenges to the present land tenure systems of the North Mkata Plain may arise as a result of land shortage rather than new land use practices. Given the present level of technology of smallholders, and the shortage of land that can be worked by hand, an increase in the smallholder population will put pressure on the land now cultivated. This pressure may lead to more intensive use of this land through the adoption of improved management techniques, some of which have been suggested above. But will this pressure lead to a desire to break up the largeholdings now held by grant? The answer is hard to predict, but already smallholder encroachment is occurring in a small way on sisal estates that have been recently abandoned or neglected. This could be a source of future conflict that could possibly be avoided by intensifying the agriculture in the present smallholder areas.

Another form of conflict that has long been a part of the character of the North Mkata Plain derives from the grazing practices of the Baraguyu. These independent-minded people seem not to recognize the sanctity of others' rights to land – be they rights of occupancy by grant or by customary law. Cattle are grazed from time to time on lands not held by the Baraguyu, especially in periods of drought in the Mkata Station-Dakawa land unit. But what lands do the Baraguyu hold rights over? Even the Baraguyu themselves have difficulty answering this question precisely. If acceptable to the Baraguyu, the establishment of range development commissions could answer the question and clarify a situation that now is a source of trouble between the Baraguyu and the farmers of the area.

Conflict also arises in the cultivated areas held by customary law because of the number of people involved in the process of land allocation. To avoid such conflict in the future, a political decision will have to be made either to vest allocation power in one person or group within a defined area of land, or to delimit the sphere of influence of the three present allocators – *jumbe, bwana shamba*, and village development committee.

CONCLUSION

Through a discussion of the land tenure systems, by assessing the general capability of the land for agriculture, and by providing maps of land use, soil associations, and land capability, this book provides data that are fundamental to the planning of agricultural development in the North Mkata Plain. This analysis of some of the important ecological and social elements that make up the agricultural system and subsystems of the area can now be coupled with economic studies and integrated into a land use planning process.

Without being excessively gloomy, one must come to the conclusion that, on the whole, prospects for development are limited owing to ecological factors. The focus in this paper has been on detailing the limiting physical features of the soils of the North Mkata Plain. But other limiting ecological factors have necessarily crept into the discussion. They have included flooding, the generally low fertility levels of the soils, the presence of tsetse fly in some areas, and the limited availability of moisture for crop development. Taken together, these ecological characteristics of the North Mkata Plain form significant obstacles to the development of agriculture.

Land tenure in the survey area has not been described as being an obstacle to all forms of development. But it is evident that the customary land tenure systems of the survey area are often associated with some undesirable land use practices. For example, a cause-and-effect relationship between land tenure and fragmentation can be claimed. And it has been maintained that many of the tenure regulations would be bottlenecks to the adoption of some of the improvements required to overcome limiting ecological factors.

Government programs that encourage Tanzanian agriculturalists to live in villages and work land communally may overcome the obstacles to improvements attributed to customary land tenure and land use practices. But it is the author's view that the legislative provisions for villagization are inadequate because of their sterile structural character. They tend toward the homogenization of African land tenure throughout Tanzania. Their inflexibility does not give recognition to the cultural diversity of the country. Nor do they recognize the advantages, both cultural and ecological, that could accrue through the incorporation of the beneficial or positive aspects of customary land tenure and land use practices.

Of course much more research is possible and necessary in the North Mkata Plain, concerned both with the agricultural development of this area and with the broader store of knowledge. A number of research topics have already been suggested in this book: detailed hydrologic analysis of the drainage systems; experimentation with new crop varieties on different soil types; close scrutiny of the economic factors of introducing various agricultural improvements; an examination of the effects of fragmentation on crop yields. For more detailed planning purposes, it would be desirable now to map the soils and the land capability of the Kilosa-Turiani land unit on a larger scale. It would also be useful to compare the tenure systems of the Ngulu with the systems of land holding practised in other areas held by customary law. These, and innumerable questions not posed, must be left to future investigations, which it is hoped will come soon for the sake of the inhabitants of this environment and of similar environments elsewhere.

Water Balance Tables for Selected Stations in or close to the North Mkata Plain

	Jan	Feb	Mar	Apr	May	June	July	Aug	Sept	Oct	Nov	Dec	Total
Station No. 96.3700: MOROGORO													
Precipitation	89	105	146	216	92	21	14	11	14	35	70	81	894
Potential evaporation	179	167	170	142	139	132	138	153	178	208	205	196	2,007
Soil moisture	0	0	0	74	27	0	0	0	0	0	0	0	0
Soil moisture change	0	0	0	+74	-47	-27	0	0	0	0	0	0	0
Actual evapotranspiration	89	105	146	142	139	48	14	11	14	35	70	81	894
Deficit	90	62	24	0	0	84	124	142	164	173	135	115	1,113
Surplus	0	0	0	0	0	0	0	0	0	0	0	0	0
Station No. 96.3710: MUSKATI MISSION													
Precipitation	125	133	169	253	118	32	21	18	28	46	79	118	1,140
Potential evaporation	148	133	155	105	142	152	161	193	223	271	209	155	2,047
Soil moisture	0	0	14	162	138	18	0	0	0	0	0	0	0
Soil moisture change	0	0	+14	+148	-24	-120	-18	0	0	0	0	0	0
Actual evapotranspiration	125	133	155	105	142	152	39	18	28	46	79	118	1,140
Deficit	23	0	0	0	0	0	122	175	195	225	130	37	907
Surplus	0	0	0	0	0	0	0	0	0	0	0	0	0
Station No. 96.3713: SCUTARI SISAL ESTATE													
Precipitation	132	147	183	182	59	5	5	6	5	27	55	106	912
Potential evaporation	181	171	170	150	135	125	131	141	167	195	206	204	1,976
Soil moisture	0	0	13	45	0	0	0	0	0	0	0	0	0
Soil moisture change	0	0	+13	+32	-45	0	0	0	0	0	0	0	0
Actual evapotranspiration	132	147	170	150	104	5	5	6	5	27	55	106	912
Deficit	49	24	0	0	31	120	126	135	162	168	151	98	1,064
Surplus	0	0	0	0	0	0	0	0	0	0	0	0	0

	Jan	Feb	Mar	Apr	May	June	July	Aug	Sept	Oct	Nov	Dec	Total
Station No. 96.3714: MARIOS SISAL ESTATE													
Precipitation	107	135	165	167	51	3	4	10	5	23	50	97	817
Potential evaporation	182	171	170	150	135	125	131	141	165	195	206	204	1,975
Soil moisture	0	0	0	17	0	0	0	0	0	0	0	0	
Soil moisture change	0	0	0	+17	-17	0	0	0	0	0	0	0	
Actual evapotranspiration	107	135	165	150	68	3	4	10	5	23	50	97	817
Deficit	75	36	5	0	67	122	127	131	160	172	156	107	1,158
Surplus	0	0	0	0	0	0	0	0	0	0	0	0	0
Station No. 96.3718: MHONDO MISSION													
Precipitation	190	142	275	350	189	51	60	43	47	78	127	228	1,780
Potential evaporation	188	179	176	151	137	128	135	145	169	194	201	201	2,004
Soil moisture	29	0	99	250	250	173	98	0	0	0	0	27	
Soil moisture change	+2	-29	+99	+151	0	-77	-75	-98	0	0	0	-27	
Actual evapotranspiration	188	171	176	151	137	128	135	141	47	78	127	201	1,680
Deficit	0	8	0	0	0	0	0	4	122	116	74	0	324
Surplus	0	0	0	48	52	0	0	0	0	0	0	0	100
Station No. 96.3719: MSOWERO													
Precipitation	133	111	158	249	89	10	8	4	5	16	49	105	937
Potential evaporation	180	169	170	146	135	128	134	146	173	202	205	199	1,987
Soil moisture	0	0	0	103	57	0	0	0	0	0	0	0	
Soil moisture change	0	0	0	+103	-46	-57	0	0	0	0	0	0	
Actual evapotranspiration	133	111	158	146	135	67	8	4	5	16	49	105	937
Deficit	47	58	12	0	0	61	126	142	168	186	156	94	1,050
Surplus	0	0	0	0	0	0	0	0	0	0	0	0	0

	Jan	Feb	Mar	Apr	May	June	July	Aug	Sept	Oct	Nov	Dec	Total
Station No. 96.3721: MVOMERO													
Precipitation	111	95	162	232	100	18	12	13	14	32	64	82	935
Potential evaporation	186	176	174	151	136	127	134	143	169	194	203	202	1,995
Soil moisture	0	0	0	81	45	0	0	0	0	0	0	0	0
Soil moisture change	0	0	0	+81	−36	−45	0	0	0	0	0	0	0
Actual evapotranspiration	111	95	162	151	136	63	12	13	14	32	64	82	935
Deficit	75	81	12	0	0	64	122	130	155	162	139	120	1,060
Surplus	0	0	0	0	0	0	0	0	0	0	0	0	0
Station No. 96.3733: MASIMBU SISAL ESTATE													
Precipitation	128	79	107	177	66	6	3	7	7	13	33	59	685
Potential evaporation	182	171	171	146	138	129	135	149	173	201	204	200	1,999
Soil moisture	0	0	0	31	0	0	0	0	0	0	0	0	0
Soil moisture change	0	0	0	+31	−31	0	0	0	0	0	0	0	0
Actual evapotranspiration	128	79	107	146	97	6	3	7	7	13	33	59	685
Deficit	54	92	64	0	41	123	132	142	166	188	171	141	1,314
Surplus	0	0	0	0	0	0	0	0	0	0	0	0	0

Soil Profile Descriptions: Morphological Data

CHALA SERIES

Locality: N of Makuyu
Latitude: S 6°17'
Longitude: E 37°25'

Parent material: Gneiss
Topography: Steeply dissected; 30% slope
Vegetation: Miombo woodland
Drainage: Excessively drained

Description at various depths

0-25 cm: Strong brown (7.5YR 5/6 moist and dry) coarse-textured soil; structureless; loose when dry; common fine, rare medium roots; stony.

25 cm: Bedrock.

DAKAWA SERIES

Locality: Dakawa
Latitude: S 6°35'
Longitude: E 37°35'

Parent material: Sandy and silty alluvium
   not recently deposited
Topography: Almost flat; 2% slope
Vegetation: Bushland-thicket
Drainage: Well-drained

Description at various depths

0-15 cm: Dark brown (7.5YR 4/2 moist) to brown (7.5YR 5/2 dry) sandy loam; weak medium granular structure; loose when dry; frequent roots.

15-30 cm: Dark yellowish brown (10YR 4/4 moist) to yellowish brown (10YR 5/4 dry) loam; weak medium subangular blocky structure; loose when dry; few medium, frequent fine roots.

30-48 cm: Very dark grayish brown (10YR 3/2 moist) to dark brown (10YR 3/3 dry) sandy loam; weak medium subangular blocky structure; loose when dry; very few very fine roots.

48–58 cm: Light brownish gray (10YR 6/2 moist) to light gray (10YR 7/2 dry) loamy sand; structureless; loose when dry; very few very fine roots.

58–100 cm: Dark reddish brown (5YR 3/2 moist and 5YR 3/3 dry) sandy loam; structureless; loose when dry; no roots.

ILONGA SERIES

Locality: Rudewa
Latitude: S 6°40'
Longitude: E 37°7'

Parent material: Sandy clay alluvium not recently deposited
Topography: Undulating; 4% slope
Vegetation: Abandoned sisal field
Drainage: Well-drained

Description at various depths

0–20 cm: Very dark gray (10YR 3/1 moist) to very dark grayish brown (10YR 3/2 dry) sandy clay loam; weak medium subangular blocky structure; hard when dry; frequent medium and fine roots.

20–50 cm: Dark brown (10YR 3/3 moist) to dark yellowish brown (10YR 3/4 dry) sandy clay loam; moderate coarse subangular blocky structure; hard when dry; common medium and frequent fine roots.

50–100 cm: Dark yellowish brown (10YR 4/4 moist) to yellowish brown (10YR 5/4 dry) clay loam; structureless; loose when dry; very few very fine roots.

KAGURU SERIES

Locality: Rudewa
Latitude: S 6°43'
Longitude: E 37°16'

Parent material: Gneiss
Topography: Undulating; 6% slope
Vegetation: Woodland
Drainage: Imperfectly drained

Description at various depths

0–15 cm: Dark brown (7.5YR 3/2 moist and dry) clay loam; weak medium granular structure; soft when dry; common medium abundant fine roots.

15–50 cm: Dark brown (10YR 3/3 moist) to dark yellowish brown (10YR 3/4 dry) clay; medium subangular blocky structure; slightly hard when dry; common fine roots; 1 to 2% Fe-concretions; few fine distinct clear mottles.

50–100 cm: Dark brown (10YR 3/3 moist) to dark yellowish brown (10YR 3/4 dry) clay; structureless and massive; hard when dry; very few fine roots; 2% Fe-concretions; common fine distinct clear mottles.

KIDETE SERIES

Locality: Kidete	Parent material: Gneiss
Latitude: S 6°20′	Topography: Rolling; 15% slope
Longitude: E 37°15′	Vegetation: Miombo woodland
	Drainage: Well-drained

Description at various depths

0–10 cm: Reddish brown (2.5YR 4/4 moist) to red (2.5YR 4/6 dry) sandy clay loam; moderate medium granular structure; slightly hard when dry; common fine roots; stony.

10–80 cm: Reddish brown (2.5YR 4/4 moist) to red (2.5YR 4/6 dry) sandy clay loam; structureless; soft when dry; friable when moist; few fine roots; very stony with stones stained with Fe and Mn.

80 cm: Bedrock.

KIDUNDA SERIES

Locality: NW of Nguru ya Ndege	Parent material: Gneiss
Latitude: S 6°35′	Topography: Gently undulating; 4% slope
Longitude: E 37°35′	Vegetation: Wooded bushland
	Drainage: Imperfectly drained

Description at various depths

0–20 cm: Very dark grayish brown (10YR 3/2 moist and dry) sandy clay loam; moderate medium granular structure; slightly hard when dry; common fine, frequent very fine roots.

20–60 cm: Dark gray (10YR 4/1 moist) to dark grayish brown (10YR 4/2 dry) sandy clay; moderate coarse angular blocky structure; very hard when dry; very few very fine roots; 1 to 2% Fe-concretions up to 10 mm in diameter.

60–100 cm: Dark reddish brown (2.5YR 3/4 moist) to reddish brown (2.5YR 4/4 dry) sandy loam; structureless; very hard when dry; 80% Fe-concretions up to 30 mm in diameter but more commonly 15 mm in diameter.

KIMALA SERIES

Locality: N of Kidete	Parent material: Gneiss
Latitude: S 6°20′	Topography: Hilly; 16% slope
Longitude: E 37°15′	Vegetation: Miombo woodland
	Drainage: Excessively drained

Description at various depths

0-10 cm: Brown (10YR 4/3 moist and 10YR 5/3 dry) medium-textured soil; moderate fine granular structure; loose when dry; common medium and fine roots; stony.

10-30 cm: Reddish brown (2.5YR 4/4 moist and dry) coarse-textured soil; structureless; loose when dry; common fine roots; very stony.

30 cm: Bedrock.

KIMAMBA SERIES

Locality: Kimamba
Latitude: S 6°47'
Longitude: E 37°8'

Parent material: Sandy alluvium not recently deposited
Topography: Flat; 1-2% slope
Vegetation: Sisal field
Drainage: Excessively drained

Description at various depths

0-50 cm: Very dark grayish brown (10YR 3/2 moist) to dark brown (10YR 3/3 dry) sandy loam; moderate medium granular structure; loose when dry; abundant coarse, medium, and fine roots.

50-100 cm: Yellowish brown (10YR 5/4 moist) to light yellowish brown (10YR 6/4 dry) sand; structureless; loose when dry; few medium and frequent fine roots.

KINGOLWIRA SERIES

Locality: NW of Nguru ya Ndege
Latitude: S 6°40'
Longitude: E 37°35'

Parent material: Gneiss and clay colluvium
Topography: Undulating; 3% slope
Vegetation: Thicket and grassland
Drainage: Well-drained

Description at various depths

0-15 cm: Dark red (2.5YR 3/6 moist) to red (2.5YR 4/6 dry) clay; moderate medium granular structure; slightly hard when dry; slightly sticky and plastic when wet; abundant fine, common medium, and very few coarse roots.

15-50 cm: Red (10R 4/6 moist and 10R 4/8 dry) clay; structureless; hard when dry; sticky and slightly plastic when wet; common fine, very few medium roots.

50-150 cm: Dark red (2.5YR 3/6 moist) to red (2.5YR 4/8 dry) clay; structureless; hard when dry; non-sticky and non-plastic when wet; very few fine roots.

KIPINDE SERIES

Locality: Makuyu
Latitude: S 6°18′
Longitude: E 37°35′

Parent material: Sandy clay colluvium
Topography: Undulating; 3% slope
Vegetation: Cotton field
Drainage: Imperfectly drained

Description at various depths

0-30 cm: Very dark grayish brown (10YR 3/2 moist) to dark brown (10YR 3/3 dry) sandy clay loam; weak fine granular structure; loose when dry; common medium, abundant fine roots.

30-60 cm: Dark brown (10YR 3/3 moist and 10YR 4/3 dry) sandy clay loam; weak fine subangular blocky structure; slightly hard when dry; few medium and fine roots.

60-110 cm: Dark yellowish brown (10YR 3/4 moist and 10YR 4/4 dry) clay loam; moderate medium subangular blocky structure to structureless; hard when dry; few fine roots; 2 to 3% Fe-concretions up to 5 mm in diameter; common fine distinct clear mottles.

KWADIHOMBO SERIES

Locality: Wakwavi Settlement
Latitude: S 6°16′
Longitude: E 37°35′

Parent material: Clay alluvium not recently deposited
Topography: Flat; 0-1% slope
Vegetation: Wooded grassland
Drainage: Imperfectly drained

Description at various depths

0-10 cm: Very dark gray (10YR 3/1 moist) to very dark grayish brown (10YR 3/2 dry) clay; weak medium subangular blocky structure; powdery between soil clods; hard when dry; few fine roots.

10-30 cm: Very dark grayish brown (10YR 3/2 moist) to dark grayish brown (10YR 4/2 dry) clay; strong coarse subangular blocky structure; hard when dry; sticky and plastic when wet; very few very fine roots; few fine distinct clear reddish to yellowish red mottles.

30-80 cm: Dark brown (7.5YR moist and 7.5YR 4/2 dry) clay; strong coarse subangular blocky structure; very hard when dry; very sticky and plastic when wet; no roots; common medium distinct clear gray mottles.

80-140 cm: Very dark gray (10YR 3/1 moist) to dark gray (10YR 4/1 dry) clay; structureless and massive; very hard when dry; very sticky and plastic when wet; no roots.

KWAVI SERIES

Locality: Wakwavi Settlement
Latitude: S 6°20'
Longitude: E 37°36'

Parent material: Clay alluvium not recently
    deposited
Topography: Flat; 1–2% slope
Vegetation: Grassland
Drainage: Poorly drained

Description at various depths

0–10 cm: Black (10YR 2/1 moist) to very dark gray (10YR 3/1 dry) sandy clay; moderate medium subangular blocky structure; firm when dry; slightly sticky and plastic when wet; frequent fine roots.

10–60 cm: Black (10YR 2/1 moist and dry) clay; strong coarse subangular blocky structure; very hard when dry; sticky and plastic when wet; few fine roots.

60–150 cm: Black (10YR 2/1 moist and dry) clay; moderate coarse angular blocky structure; very hard when dry; sticky and plastic when wet; many medium distinct clear mottles; 2% carbonate concretions less than 2 cm in diameter.

MAGOLE SERIES

Locality: Magole
Latitude: S 6°23'
Longitude: E 37°27'

Parent material: Sandy and silty alluvial
    clays not recently deposited
Topography: Almost flat; 2% slope
Vegetation: Sunflower field
Drainage: Well-drained

Description at various depths (from Johnson and Tiarks, 1969)

0–20 cm: Black (10YR 2/1 moist) to very dark brown (10YR 2/2 dry) loam; weak coarse subangular blocky structure; extremely hard when dry; sticky and plastic when wet; few fine roots.

20–33 cm: Very dark brown (10YR 2/2 moist) to very dark, grayish brown (10YR 3/2 dry) loam; moderate coarse subangular blocky structure; extremely hard when dry; sticky and plastic when wet; few fine roots.

33–61cm: Black (10YR 2/1 moist) to very dark grayish brown (10YR 3/2 dry) sandy loam; moderate medium subangular blocky structure; firm when moist; slightly sticky and plastic when wet; few fine roots.

61–81cm: Dark yellowish brown (10YR 3/4 moist) to reddish brown (5YR 4/4 dry) loam; weak medium subangular blocky structure; very friable when moist; slightly sticky and plastic when wet; few fine roots.

81-118 cm: Reddish brown (5YR 4/4 moist) to yellowish red (5YR 4/8 dry) loam; weak medium subangular blocky structure; very friable when moist; non-sticky and non-plastic when wet; no roots; frequent mica fragments less than 2 mm across.

MAKUYU SERIES

Locality: Makuyu   Parent material: Sandy clay colluvium
Latitude: S 6°18'   Topography: Undulating; 3% slope
Longitude: E 37°25'   Vegetation: Maize field
                                          Drainage: Well-drained

Description at various depths

0-15 cm: Dark brown (7.5YR 3/2 moist and 7.5YR 4/4 dry) clay loam; moderate medium granular structure; very hard when dry; common medium and fine roots.

15-55 cm: Dark reddish brown (5YR 3/2 moist) to dark reddish gray (5YR 4/2 dry) sandy clay loam; moderate coarse subangular blocky structure; very hard when dry; common fine and very fine roots; 1 to 2% Fe-concretions of less than 5 mm in diameter.

55-105 cm: Dark brown (7.5YR 3/2 moist and 7.5YR 4/2 dry) clay loam; moderate medium subangular blocky structure to structureless; very hard when dry; few very fine roots; 1 to 2% Fe-concretions of less than 5 mm in diameter.

MKATA SERIES

Locality: Mkata Station   Parent material: Sandy clay alluvium not
Latitude: S 6°46'   recently deposited
Longitude: E 37°47'   Topography: Flat; 0-1% slope
                                          Vegetation: Wooded grassland
                                          Drainage: Imperfectly drained

Description at various depths (from Johnson and Tiarks, 1969)

0-60 cm: Black (10YR 2/1 moist) to dark gray (10YR 4/1 dry) sandy clay; strong coarse subangular blocky structure; very hard when dry; sticky and plastic when wet; common fine, few medium roots.

60-97 cm: Black (10YR 2/1 moist and dry) sandy clay; moderate coarse subangular blocky structure; very hard when dry; sticky and plastic when wet; few fine roots; 15% carbonate concretions up to 5 cm in diameter.

97-152 cm: Black (10YR 2/1 moist and dry) sandy clay; structureless and massive; very hard when dry; sticky and plastic when wet; no roots; 10% carbonate concretions up to 5 cm in diameter.

MKUNDI SERIES

Locality: Magole

Latitude: S 6°20'

Longitude: E 37°20'

Parent material: Sandy alluvium recently deposited

Topography: Undulating; 3% slope

Vegetation: Riverine woodland

Drainage: Excessively drained

Description at various depths

0-10 cm: Very dark grayish brown (10YR 3/2 moist) to dark grayish brown (10YR 4/2 dry) sandy loam; weak medium granular structure; slightly hard when dry; common fine and medium roots.

10-35 cm: Brown (10YR 5/3 moist) to light yellowish brown (10YR 6/4 dry) sand; structureless; slightly hard when dry; common fine and few medium roots.

35-55 cm: Very dark grayish brown (10YR 3/2 moist) to dark grayish brown (10YR 4/2 dry) loamy sand; structureless; hard when dry; very few very fine roots.

55-100 cm: Dark yellowish brown (10YR 4/4 moist) to yellowish brown (10YR 5/4 dry) loamy sand; structureless; very hard when dry; very few very fine roots.

MOROGORO SERIES

Locality: Masimbu

Latitude: S 6°42'

Longitude: E 37°35'

Parent material: Gneissic drift

Topography: Undulating; 4% slope

Vegetation: Wooded bushland

Drainage: Well-drained

Description at various depths

0-30 cm: Dark reddish brown (5YR 3/3 moist and 5YR 3/4 dry) sandy loam; moderate medium granular structure; soft to slightly hard when dry; very few very fine roots.

30-150 cm: Dark red (2.5YR 3/6 moist and dry) sandy loam; structureless; very friable when moist; soft when dry; very few very fine roots; small amounts of quartz fragments; numerous termite channels.

MSOWERO SERIES

Locality: Msowero
Latitude: S 6°32'
Longitude: E 37°13'

Parent material: Sandy and silty clay
   alluvium recently deposited
Topography: Flat; 0–2% slope
Vegetation: Sugarcane field
Drainage: Poorly drained

Description at various depths

0–20 cm: Very dark grayish brown (10YR 3/2 moist and dry) clay loam; moderate medium angular blocky structure; extremely hard when dry; common fine roots; common distinct reddish brown mottles.

20–50 cm: Very dark grayish brown (10YR 3/2 moist and dry) clay loam; moderate coarse subangular blocky structure to structureless; hard when dry; few fine roots; common clear distinct dark mottles.

50–150 cm: Black (10YR 2/1 moist) to very dark grayish brown (10YR 3/2 dry) sandy clay loam; structureless and massive; very hard when dry; slightly sticky and slightly plastic when wet; mottled throughout; 2 to 3% carbonate concretions up to 2 mm in diameter.

MVOMERO SERIES

Locality: Mvomero
Latitude: S 6°17'
Longitude: E 37°26'

Parent material: Sandy clay alluvium
   recently deposited
Topography: Undulating; 2% slope
Vegetation: Riverine grassland and
   bushland
Drainage: Imperfectly drained

Description at various depths

0–30 cm: Very dark grayish brown (10YR 3/2 moist and dry) sandy clay loam; moderate medium subangular blocky structure; very hard when dry; common fine and medium roots.

30–60 cm: Very dark grayish brown (10YR 3/2 moist) to dark brown (10YR 3/1 dry) sandy clay loam; moderate medium angular blocky structure; hard when dry; few fine and very fine roots.

60–100 cm: Dark yellowish brown (10YR 3/4 moist and 10YR 4/4 dry) sandy clay loam; structureless and massive; very hard when dry; slightly sticky and plastic when wet; no roots; moderate clear distinct mottles.

MVUMI SERIES

Locality: Mvumi
Latitude: S 6°35'
Longitude: E 37°11'

Parent material: Sandy clay alluvium not
   recently deposited
Topography: Flat; 1-2% slope
Vegetation: Grassland
Drainage: Poorly drained

Description at various depths

0-15 cm: Dark reddish brown (5YR 3/3 moist) to reddish brown (5YR 4/3 dry) loam; strong medium crumb structure; loose when dry; few medium, frequent fine and very fine roots; 2 to 3% Fe-concretions up to 2 mm in diameter.

15-50 cm: Dark brown (7.5YR 4/2 moist and 7.5YR 3/2 dry) loam; strong coarse columnar structure; extremely hard when dry; few very fine roots; 7 to 10% concretions of mixed iron and carbonate composition.

50-90 cm: Very dark grayish brown (10YR 3/2 moist) to dark grayish brown (10YR 4/2 dry) silt loam; structureless and massive; firm when moist; very few very fine roots; 10% mixed iron and carbonate concretions; common medium distinct mottles.

90-140 cm: Dark reddish brown (5YR 3/2 moist and 5YR 2/2 dry) clay loam; structureless; very firm when moist; no roots; abundant Fe-concretions; abundant coarse distinct clear mottles.

NDEGE SERIES

Locality: Nguru ya Ndege
Latitude: S 6°41'
Longitude: E 37°37'

Parent material: Gneiss
Topography: Dissected; 30% slope
Vegetation: Miombo woodland
Drainage: Excessively drained

Description at various depths

0-10 cm: Light yellowish brown (10YR 6/4 moist) to brownish yellow (10YR 6/6 dry) coarse-textured soil; weak fine granular structure; loose when dry; few fine and medium roots.

10-25 cm: Yellowish brown (10YR 5/4 moist and 10YR 5/6 dry) coarse-textured soil; structureless; loose when dry; few fine roots; stony.

25 cm: Bedrock.

VILANZA SERIES

Locality: Masimbu　　　　　　　Parent material: Gneissic drift
Latitude: S 6°44'　　　　　　　　Topography: Undulating; 3% slope
Longitude: E 37°33'　　　　　　　Vegetation: Wooded grassland
　　　　　　　　　　　　　　　　Drainage: Well-drained

Description at various depths

0-15 cm: Reddish brown (5YR 4/4 moist) to yellowish red (5YR 4/6 dry) sandy clay; weak medium granular structure; slightly hard when dry; friable when moist; many fine common medium roots.

15-60 cm: Reddish brown (5YR 4/4 moist) to yellowish red (5YR 4/6 dry) clay; moderate medium granular structure to structureless; slightly hard when dry; very friable when moist; few fine and very fine roots.

60-125 cm: Reddish brown (2.5YR 4/4 moist) to red (2.5YR 4/6 dry) clay; structureless; loose when dry; very friable when moist; very few fine roots; 1 to 2% carbonate concretions.

WAMI SERIES

Locality: Wakwavi Settlement　　Parent material: Clay alluvium recently
Latitude: S 6°22'　　　　　　　　　deposited
Longitude: E 37°37'　　　　　　　Topography: Flat; 0-1% slope
　　　　　　　　　　　　　　　　Vegetation: Grassland
　　　　　　　　　　　　　　　　Drainage: Very poorly drained

Description at various depths

0-15 cm: Dark gray (10YR 4/1 moist and dry) clay; weak fine granular structure; loose when dry; friable when moist; common fine, few medium roots; slight mottling.

15-45 cm: Gray (10YR 5/1 moist) to dark gray (10YR 4/1 dry) clay; structureless and massive; very hard when dry; firm when moist; very few very fine roots; many medium distinct clear mottles.

45-100 cm: Dark gray (10YR 4/1 moist) clay; structureless and massive; sticky and plastic when wet; very few very fine roots; many coarse distinct clear mottles.

Soil Profile Descriptions: Analytical Data

Soil associate	Depth (cm)	pH in water 1:5	Total N (%)	Organic C (%)	C:N Ratio	Total P (ppm)	K	Na	Mg	Ca	Mn	CEC	Sat. (%)	Clay (%)	Silt (%)	Sand (%)
Dakawa	0– 15	5.6	0.13	4.11	31.6	2	1.01	0.19	4.20	2.50	0.06	11.55	68	4	46	50
	15– 30	5.9	0.11	0.61	5.5	1	0.74	0.19	7.30	2.42	0.02	15.24	70	12	40	48
	30– 48	6.2	0.11	0.46	4.1	11	0.57	0.18	5.33	3.20	0.02	11.63	80	18	30	52
	48– 58	6.6	0.09	0.36	4.0	n.d.	0.42	0.24	4.33	3.75	0.03	10.95	80	2	18	80
	58–100	7.0	0.8	0.25	3.1	n.d.	0.36	0.26	3.76	5.61	0.26	13.48	76	16	12	72
Ilonga	0– 10	6.8	0.21	n.d.	n.d.	270	1.62	0.22	6.00	18.90	0.03	29.80	90	n.d.	n.d.	n.d.
	10– 20	7.0	0.18	n.d.	n.d.	265	1.17	0.24	6.40	18.80	0.01	28.50	93	n.d.	n.d.	n.d.
	20– 40	7.0	0.14	n.d.	n.d.	110	1.06	0.25	6.60	15.90	0.01	26.00	92	n.d.	n.d.	n.d.
	40– 60	7.0	0.09	n.d.	n.d.	49	0.62	0.15	4.90	9.00	0.01	16.40	90	n.d.	n.d.	n.d.
	60–100	6.9	0.07	n.d.	n.d.	63	0.45	0.10	2.90	5.90	0.01	10.80	87	n.d.	n.d.	n.d.
Kaguru	0– 15	6.3	0.06	0.41	6.8	120	7.64	0.16	7.88	7.81	0.10	35.74	66	40	38	22
	15– 50	6.3	0.10	1.02	10.2	110	6.57	0.15	11.23	13.75	0.10	37.41	85	52	27	21
	50–100	6.3	0.17	1.36	8.0	80	4.63	0.09	9.63	11.88	0.08	29.89	88	54	24	22
Kidete	0– 10	7.2	0.07	1.52	21.7	2	3.24	0.11	9.58	9.06	0.04	31.47	70	32	21	47
	10– 80	6.1	0.24	0.37	1.5	1	0.38	0.12	13.57	13.75	0.04	33.97	82	20	21	59
Kidunda	0– 20	6.0	0.21	0.89	4.2	2	1.81	0.31	6.17	4.69	0.04	22.44	58	23	13	64
	20– 60	7.5	0.18	0.84	4.7	1	1.47	0.26	7.58	20.32	0.05	34.91	85	41	3	56
	60–100	n.d.	n.d.	n.d.	n.d.	n.d.	n.d.	n.d.	n.d.	n.d.	n.d.	n.d.	n.d.	n.d.	n.d.	n.d.
Kimamba	0– 50	5.2	0.18	0.65	3.6	25	0.88	9.18	3.83	2.81	0.04	12.03	56	11	11	78
	50–100	6.5	0.27	0.26	1.0	13	0.46	0.26	1.25	1.09	0.22	6.04	51	3	4	93
Kingolwira	0– 15	5.7	0.13	1.20	9.2	70	1.60	0.10	0.80	1.60	0.15	9.55	55	49	15	36
	15– 50	5.1	0.07	0.50	7.1	32	0.44	0.10	2.20	0.57	0.04	7.98	42	60	3	37
	50–150	5.1	0.06	0.29	5.0	29	0.56	0.13	2.72	0.44	0.39	8.48	50	61	4	35

Soil associate	Depth (cm)	pH in water 1:5	Total N (%)	Organic C (%)	C:N Ratio	Total P (ppm)	Extractable cations and total cation exchange capacity (me/100 g soil)								Particle size		
							K	Na	Mg	Ca	Mn	CEC	Sat. (%)	Clay (%)	Silt (%)	Sand (%)
Kipinde	0– 30	7.9	0.17	2.31	13.6	160	2.05	0.11	16.38	9.76	0.03	30.46	93	25	24	41
	30– 60	7.7	0.27	2.29	8.5	88	0.68	0.13	6.06	12.90	0.01	21.26	93	30	15	55
	60–100	8.0	0.11	0.63	5.7	24	0.50	0.38	6.56	14.65	0.00	24.82	89	31	18	51
Kwudihombo	0– 10	7.9	0.14	1.26	9.0	16	1.64	1.01	18.06	22.17	0.01	42.89	100	48	11	41
	10– 30	8.0	0.10	0.60	6.0	7	0.78	1.21	16.53	22.28	0.02	40.82	100	48	10	42
	30– 80	8.1	0.09	0.43	4.8	2	0.77	14.63	16.66	23.11	0.01	57.47	96	51	12	37
	80–140	8.4	0.02	0.22	11.0	2	0.91	20.12	24.12	24.09	0.01	70.66	98	45	14	45
Kwavi	0– 10	6.5	0.08	0.65	8.1	21	2.04	0.12	29.17	14.07	0.04	45.89	99	47	16	37
	10– 60	7.4	0.24	0.37	1.5	8	1.68	2.31	26.25	20.94	0.02	52.24	98	54	14	32
	60–150	8.1	0.14	0.02	0.1	2	2.40	19.16	27.71	17.50	0.01	66.78	100	49	14	37
Magolo†	0– 20	6.6	0.16	1.82	11.4	n.d.	1.08	0.21	2.90	17.30	0.07	23.40	92	26	35	38
	20– 33	6.9	0.13	1.39	10.7	n.d.	0.95	0.22	2.70	17.20	0.02	n.d.	n.d.	24	34	41
	33– 61	7.1	n.d.	0.50	n.d.	n.d.	0.56	0.27	2.00	7.90	0.02	n.d.	n.d.	16	20	63
	81–118	7.0	n.d.	n.d.	n.d.	n.d.	0.61	0.24	4.40	10.40	0.02	n.d.	n.d.	20	35	44
Makuyu	0– 15	6.7	0.04	0.49	12.3	60	5.05	0.24	12.98	10.63	0.04	32.15	90	31	18	51
	15– 55	6.7	0.17	0.91	5.4	24	1.73	0.15	14.58	3.75	0.01	21.97	92	29	14	57
	55–105	6.7	0.11	0.13	1.2	11	1.05	0.08	8.17	5.00	0.01	15.38	93	31	10	59
Mkata†	0– 60	8.1	0.07	0.76	10.9	0.10	0.74	1.00	8.00	25.70	0.02	35.46	100	44	10	46
	60– 97	8.1	0.04	0.38	9.5	8	0.86	7.65	16.40	25.01	0.03	39.35	100	46	18	36
	97–152	8.3	0.03	0.29	9.7	2	0.74	30.50	33.50	30.00	0.02	94.76	100	48	10	42
Mkundi	0– 10	4.8	0.14	1.89	13.5	n.d.	0.45	0.08	0.91	3.41	0.14	13.83	36	18	8	74
	10– 35	4.6	0.05	1.	22.4	n.d.	0.17	0.12	0.43	2.03	0.10	7.50	38	7	8	85
	35– 55	5.4	0.05	0.62	12.4	n.d.	0.11	0.07	0.44	2.06	0.03	4.53	60	13	6	81
	55–100	6.6	0.01	0.08	8.0	n.d.	0.11	0.16	0.16	1.08	0.02	3.00	51	12	8	80

Soil associate	Depth (cm)	pH in water 1:5	Total N (%)	Organic C (%)	C:N Ratio	Total P (ppm)	Extractable cations and total cation exchange capacity (me/100 g soil)						Sat. (%)	Particle size		
							K	Na	Mg	Ca	Mn	CEC		Clay (%)	Silt (%)	Sand (%)
Morogoro	0–30	5.7	0.08	2.26	28.3	435	0.64	0.48	0.34	1.74	0.00	15.20	21	10	18	72
	30–150	4.8	0.05	1.02	20.4	275	0.14	0.48	0.51	6.64	0.00	22.07	35	14	24	62
Msowero	0–20	6.8	0.20	2.11	10.6	n.d.	0.54	0.48	6.21	12.18	0.11	27.89	70	38	30	32
	20–50	6.9	0.06	0.92	15.3	n.d.	0.29	0.48	4.18	8.13	0.04	22.62	58	29	32	39
	50–150	7.4	0.02	0.18	9.0	n.d.	0.28	0.69	4.29	13.22	0.04	31.93	58	31	22	47
Mvomero	0–30	6.5	0.13	2.18	16.8	n.d.	0.64	0.12	5.39	13.06	0.08	24.73	78	30	22	48
	30–60	6.2	0.05	0.65	13.0	n.d.	0.51	0.40	3.65	6.21	0.04	14.22	76	29	18	53
	60–100	6.6	0.02	0.28	14.0	n.d.	0.22	0.22	2.61	3.00	0.03	6.83	89	28	19	53
Mvumi	0–15	6.9	0.28	3.87	13.8	580	0.86	0.29	2.61	3.17	0.05	15.58	44	12	40	48
	15–60	5.5	0.02	1.28	64.0	235	0.26	6.32	0.55	3.53	0.10	20.52	52	16	32	52
	50–90	7.5	0.02	0.64	32.0	340	0.45	15.87	0.73	4.06	0.00	32.11	65	4	50	46
	90–140	5.5	0.02	0.58	29.0	185	0.55	17.94	0.94	4.30	0.06	33.69	71	20	18	62
Vilanza	0–15	5.8	0.18	0.78	4.3	1	1.47	0.41	5.67	3.88	0.03	28.65	40	45	10	45
	15–60	5.2	0.18	0.26	1.4	1	0.69	0.41	6.75	3.75	0.01	20.73	56	53	10	37
	60–125	5.6	0.11	0.31	2.8	1	0.28	0.61	5.62	3.18	0.12	23.35	42	56	8	36
Wami	0–15	6.8	0.31	4.12	13.3	n.d.	0.66	0.49	6.01	12.06	0.08	27.57	70	40	31	29
	15–45	6.8	0.13	2.33	17.9	n.d.	0.11	0.36	4.18	12.02	0.06	28.84	58	45	28	27
	45–100	7.2	0.03	0.65	21.7	n.d.	0.13	0.88	4.28	6.03	0.12	19.39	59	46	21	33

* Lock, 1969.

† Johnson and Tiarks, 1969.

n.d. Values not determined or not available.

Selected Distinguishing Characteristics of the Soil Series of the North Mkata Plain

VERTISOLS OF TOPOGRAPHIC DEPRESSIONS

	Proportion of ferric oxide mottling	Tendency towards surface mulching and gilgai
Mkata	Low	High
Kwavi	Low-medium	Medium
Kwadihombo	Medium	Low

EUTROPHIC BROWN SOILS

	Subsoil accumulation of iron concretions	Drainage	Moist colour of subsoil
Ilonga	None	Well-drained	Dark brown (10YR 3/3) to dark yellowish brown (10YR 3/4 and 4/4)
Kipinde	Slight	Imperfectly drained	Dark brown (10YR 3/3) to dark yellowish brown (10YR 3/4 and 4/4)
Makuyu	Moderate	Well-drained	Dark reddish brown (5YR 3/2) to dark brown (7.5YR 3/2)

FERRALLITIC SOILS, FERRISOLS, AND FERRUGINOUS TROPICAL SOILS

	CEC	Weatherable materials	Drainage
Vilanza	> 20 me/100 g	High	Well-drained, deep
Morogoro	> 20 me/100 g	Medium	Well-drained, deep
Kingolwira	< 20 me/100 g	Low	Well-drained, deep
Kidete	> 20 me/100 g	High	Well-drained, moderately shallow
Kidunda	> 20 me/100 g	High	Imperfectly drained
Kaguru	> 20 me/100 g	Medium	Imperfectly drained

WEAKLY DEVELOPED SOILS ON LOOSE SEDIMENTS NOT RECENTLY DEPOSITED

	CEC of subsoil	Sand content of subsoil
Magole	> 20 me/100 g	< 50%
Dakawa	< 20 me/100 g	50–80%
Kimamba	< 20 me/100 g	> 80%

WEAKLY DEVELOPED SOILS ON LOOSE SEDIMENTS RECENTLY DEPOSITED

	CEC of subsoil	Sand content of subsoil
Msowero	> 20 me/100 g	< 50%
Wami	> 20 me/100 g	< 50%
Mvomero	< 20 me/100 g	50–80%
Mkundi	< 20 me/100 g	> 80%

LITHOSOLS

	Colour below 10 cm	Texture below 10 cm
Ndege	Yellowish brown (10YR 5/4)	Coarse
Chala	Strong brown (7.5YR 5/6)	Coarse
Kimala	Reddish brown (2.5YR 4/4)	Medium

Bibliography

Aandahl, A.R. 1958. 'Soil survey interpretation – theory and purpose,'
Proceedings, Soil Science Society of America 22: 152-4

Ackerman, E.A. 1958. *Geography as a Fundamental Research Discipline.*
Chicago: University of Chicago Press

Ad Hoc Committee on Geography. 1956. *The Science of Geography.*
Washington: National Academy of Science – National Research Council

Akehurst, B.C. and A. Sreedharan. 1965. 'Time of planting – a brief review of
experimental work in Tanganyika 1956-62,' *East African Agricultural and
Forestry Journal* 30: 189-201

Allison, L.E. 1965. 'Organic carbon.' In *Methods of Soil Analysis,* ed. C.A. Black
et al. Madison: American Society of Agronomy, Monograph No. 9

Anderson, B. 1961. *The Rufiji Basin, Tanganyika. VII. Soils of the Main Irrigable
Areas.* Dar es Salaam: FAO.

- 1963. *Soils of Tanganyika.* Bulletin No. 16, Ministry of Agriculture. Dar es
Salaam: Government Printer

- 1967a. 'Soils.' In *Atlas of Tanzania.* Dar es Salaam: Government Printer

- 1967c. 'Some characteristics of the soils.' In *Atlas of Tanzania.* Dar es Salaam:
Government Printer

Anderson, G.D. 1956. 'Response of maize to application of compound fertilizers
on farmers' fields in ten districts of Tanzania.' *East African Agricultural and
Forestry Journal* 34: 382-97

Anon. 1916. *Extracts from German Ordinances and Decrees of German East
Africa.* Nairobi: Standard Printing

- 1956. 'The East Africa Royal Commission and African land tenure.' *Journal
of African Administration* 8: 69-74

- 1965. *The Canada Land Inventory: Soil Capability Classification for
Agriculture.* Ottawa: Queen's Printer

Attems, M. 1968. 'Permanent cropping in the Usambara Mountains: The relevancy of the minimum benefit thesis.' In *Smallholder Farming and Smallholder Development in Tanzania*, ed. H. Ruthenberg. Munich: Weltforum Verlag

Avery, B.W. 1969. 'Problems of soil classification.' In *The Soil Ecosystem*, ed. J.G. Sheals. London: Systematics Association

Baker, A.R.H. 1973. 'Adjustments to distance between farmstead and field: Some findings from the southwestern Paris Basin in the early nineteenth century.' *The Canadian Geographer* 17: 259-75

Baker, R.M. 1970. *The Soils of Tanzania.* Dar es Salaam: FAO (mimeo)

Barnes, C.P. 1949. 'Interpretive soil classification: Relation to purpose.' *Soil Science* 67: 127-9

Beidelman, T.O. 1960. 'The Baraguyu.' *Tanganyika Notes and Records* 55: 244-78

- 1961. 'A note on Baraguyu house-types and Baraguyu economy.' *Tanganyika Notes and Records* 56: 56-66

- 1962a. 'A history of Ukaguru: 1857-1916.' *Tanganyika Notes and Records* 58/59: 11-39

- 1962b. 'Demographic map of the Baraguyu.' *Tanganyika Notes and Records* 58/59: 8-10

- 1967. *The Matrilineal Peoples of Eastern Tanganyika.* London: International African Institute

Berry, L. 1968. 'Problems of land use planning.' *East African Agricultural and Forestry Journal* 33: 46-50

- 1969. 'Physical features.' In *East Africa: Its Peoples and Resources*, ed. W.T.W. Morgan. Nairobi: Oxford University Press

- 1971a. 'An outline of information needs in soil and land potential analysis.' *Proceedings, Second Conference on Land Use in Tanzania.* Morogoro: University of Dar es Salaam (mimeo)

- ed. 1971b. *Tanzania in Maps.* London: University of London Press

Berry, L. and E. Berry. 1969a. *A Preliminary Sub-division of Districts into Rural Economic Zones: A Map with Key.* Research Notes No. 4. Dar es Salaam: BRALUP

- 1969b. *Land Use in Tanzania by District.* Research Notes No. 6. Dar es Salaam: BRALUP

Berry, L., D. Conyers, and J. McKay. 1969. 'Regional and rural planning — practice and possibilities in Tanzania.' *Proceedings, Conference on Urban and Regional Planning in National Development.* Kampala: Makerere College Institute of Social Research (mimeo)

- 1970. 'Scale and planning in Tanzania.' *Proceedings, East Africa Social Science Conference.* Dar es Salaam: University of Dar es Salaam (mimeo)

Berry, L., J.D. Heijnen, and J.R. Pitblado. 1970. 'River basin and large areas surveys.' In *Water Development – Tanzania: A Critical Review of Research*, Research Paper No. 12. Dar es Salaam: BRALUP

Biebuyck, D.P., ed. 1966. *African Agrarian Systems.* London: Oxford University Press

- 1968. 'Land tenure: Introduction.' *International Encyclopedia of Social Sciences* 8: 562-6

Boesen, J. 1976. 'Tanzania from ujamaa to villagization.' *Proceedings, Development in Tanzania since 1967 Conference.* Toronto: University of Toronto

Bolton, A. 1971a. 'Territorial maize variety trails in Tanzania 1966-70.' *East African Agricultural and Forestry Journal* 37: 109-24

- 1971b. 'Response of maize varieties in Tanzania to different plant populations and fertilizer levels.' *Experimental Agriculture* 7: 193-203

Boserup, E. 1965. *The Conditions of Agricultural Growth.* Chicago: Aldine

Boyer, J. 1972. 'Soil potassium.' In *Soils of the Humid Tropics*, ed. Committee on Tropical Soils. Washington: National Academy of Sciences

Brain, J.L. 1962. 'The Kwere of the Eastern Province.' *Tanganyika Notes and Records* 58/59: 231-41

Bremner, J.M. 1965. 'Total nitrogen.' In *Methods of Soil Analysis*, ed. C.A. Black et al. Madison: American Society of Agronomy, Monograph No. 9

Bridges, E.M. 1970. *World Soils.* London: Cambridge University Press

Brock, B. 1969. 'Customary land tenure, "individualization," and agricultural development in Uganda.' *East African Journal of Rural Development* 2: 1-27

Brookfield, H.C. 1973. 'On one geography and a Third World.' *Transactions, Institute of British Geographers* 58: 1-19

Buckman, H.O. and N.C. Brady. 1969. *The Nature and Properties of Soils.* New York: Macmillan

BRALUP. 1973. *Annual Report 1972/73.* Dar es Salaam: BRALUP

Buringh, P. 1970. *Introduction to the Study of Soils in Tropical and Subtropical Regions.* Wageningen: Centre for Agricultural Publishing and Documentation

Burtt, B.D. 1942. 'Some East African vegetation communities.' *Journal of Ecology* 30: 65-146

Calton, W.E. 1954. 'An experimental pedological map of Tanganyika.' *Proceedings, Second Inter-African Soils Conference* 1: 237-40

- 1959. 'Generalizations on some Tanganyika soil data.' *Journal of Soil Science* 10: 169-76

Chang, J-H. 1968. *Climate and Agriculture*. Chicago: Aldine

Chisholm, M. 1962. *Rural Settlement and Land Use*. London: Hutchinson

Collinson, M.P. 1963. *Farm Management Survey Report No. 2*. Ukiriguru (Tanzania): Western Research Station (mimeo)

Connell, J. 1971. 'The geography of development.' *Area* 3: 123-8

Dagg, M. 1965. 'A rational approach to the selection of crops for areas of marginal rainfall in East Africa.' *East African Agricultural and Forestry Journal* 30: 296-300

– 1969. 'Water requirements of crops.' In *East Africa: Its Peoples and Resources*, ed. W.T.W. Morgan. Nairobi: Oxford University Press

De Lisle, D. de G. 1978. 'Effects of distance internal to the farm: a challenging subject for North American geographers.' *The Professional Geographer* 30: 278-88

D'Hoore, J.L. 1964. *Soil Map of Africa: Scale 1 to 5,000,000: Explanatory Monograph*. Joint Project No. 11. Lagos: Commission for Technical Co-operation in Africa

– 1968. 'The classification of tropical soils.' In *The Soil Resources of Tropical Africa*, ed. R.P. Moss. Cambridge: Cambridge University Press

Dias, J.S., J.C. Povoas, and J.B. Macedo. 1959. 'A note on the origin of the colour of tropical black clay soils (gravingra).' *Proceedings, Third Inter-African Soils Conference* 1: 159-63

Dobson, E.B. 1954. 'Comparative land tenure of ten Tanganyika tribes.' *Journal of African Administration* 6: 80-91

Dolfi, D. 1963. *Water Resources Potential of the Wami Basin*. Rome: FAO

Dudal, R. 1965. *Dark Clay Soils of Tropical and Subtropical Regions*. Rome: FAO

Duthie, D.W. 1956. 'Soils.' In *Atlas of Tanganyika* 2nd ed. Dar es Salaam: Government Printer

East Africa Royal Commission. 1955. *East Africa Royal Commission 1953-1955 Report*. London: HMSO

Eicher, C. and L. Witt, eds. 1964. *Agriculture in Economic Development*. New York: McGraw-Hill

Elias, T.O. 1956. *The Nature of African Customary Law*. Manchester: Manchester University Press

Ellman, A.O. 1970. 'Progress, problems, and prospects in ujamaa development in Tanzania. *Proceedings, East African Agricultural Economics Society Conference*. Dar es Salaam: University of Dar es Salaam (mimeo)

FAO 1969. *Guidelines for Soil Profile Description*. Rome: FAO

FitzPatrick, E.A. 1971. *Pedology*. Edinburgh: Oliver and Boyd

Flach, K.W. 1963. 'Soil investigations and the 7th Approximation.' *Proceedings, Soil Science Society of America* 27: 226-8

Floyd, B. 1969. 'Toward a more specific geography of traditional agriculture in the tropics: or good-bye to machete and dibble stick.' *Professional Geographer* 21: 248-51

Found, W.C. 1970. 'Towards a general theory relating distance between farm and home to agricultural production.' *Geographical Analysis* 2: 165-76

Fozzard, P.M.H. 1965. *Kimamba Quarter Degree Sheet No. 183.* Tanzania Geological Survey. Dodoma: Government Printer

Fraser-Smith, W.S. 1965. 'Agricultural transformation through the Village Settlement Scheme.' In *Agricultural Development in Tanzania*, ed. H.E. Smith. Dar es Salaam: Institute of Public Administration

Fuggles-Couchman, N.R. 1964. *Agricultural Change in Tanganyika 1945-60.* Stanford: Stanford University Press

Furon, R. 1963. *The Geology of Africa.* Edinburgh: Oliver and Boyd

Gaitskell, A. 1959. *Report on Land Tenure and Land Use Problems in the Trust Territories of Tanganyika and Ruanda-Urundi.* FAO/59/2/951. Rome: FAO

German East Africa. 1903. *Circular Regarding Lease of Crown Lands* (Pamphlets Section, Library, University of Dar es Salaam)

Gilbert, A. 1971. 'Some thoughts on the "New Geography" and the study of development.' *Area* 3: 123-8

Ginsburg, N. 1973. 'From colonialism to national development: Geographical perspectives on patterns and policies.' *Annals, Association of American Geographers* 63: 1-21

Gould, P.R. 1969/70. 'Tanzania 1920-63: The spatial impress of the modernization process.' *World Politics* 22: 149-70

Gregor, H.F. 1970. *Geography of Agriculture: Themes in Research.* Englewood Cliffs: Prentice-Hall

Gulliver, P.H. 1959. 'A tribal map of Tanganyika.' *Tanganyika Notes and Record* 52: 61-74

Gwassa, G.C.K. 1969. 'The German intervention and African resistance in Tanzania.' In *A History of Tanzania*, eds. I.N. Kimambo and A.J. Temu. Nairobi: East African Publishing House

Hailey, Lord. 1952. 'The land tenure problem in Africa.' *Journal of African Administration* 4: 3-7

Hardy, F. 1954. 'Some ecological aspects of tropical pedology.' *Journal of Tropical Geography* 2: 1-8

Hartshorne, R. 1939. *The Nature of Geography.* Lancaster, Pa.: Association of American Geographers

Harvey, D. 1969. *Explanation in Geography.* London: Edward Arnold

Haughton, S.H. 1963. *The Stratigraphic History of Africa South of the Sahara.* New York: Hafner

Hayami, Y. and V.W. Ruttan. 1971. *Agricultural Development: An International Perspective.* Baltimore: Johns Hopkins University Press

Heijnen, J.D. 1969. *Mechanised Block Cultivation Schemes in Mwanza Region.* Research Paper No. 9. Dar es Salaam: BRALUP

Helleiner, G.K., ed. 1968. *Agricultural Planning in East Africa.* Nairobi: East African Publishing House

Hill, J.F.R. and J.P. Moffett. 1955. *Tanganyika: A Review of Its Resources and Their Development.* Dar es Salaam: Government Printer

Hill, P. 1963. *The Migrant Cocoa Farmers of Southern Ghana: A Study in Rural Capitalism.* Cambridge University Press

Hirst, M.A. 1970. 'Rural settlement and land use: A note on Tanzania.' *Professional Geographer* 5: 258-9

Hodder, B.W. 1968. *Economic Development in the Tropics.* London: Methuen

Hodder, B.W. and D.R. Harris, eds. 1967. *Africa in Transition.* London: Methuen

Igbozurike, M.U. 1970. 'Fragmentation in tropical agriculture: An overrated phenomenon.' *Professional Geographer* 22: 321-5

Iliffe, J. 1969. *Tanganyika under German Rule, 1905-1912.* Cambridge University Press

International Bank for Reconstruction and Development. 1961. *The Economic Development of Tanganyika.* Baltimore: Johns Hopkins University Press

Jackson, I.J. 1970. *Rainfall over the Ruvu Basin and Surrounding Area.* Research Report No. 9. Dar es Salaam: BRALUP

Jackson, M.L. 1964. 'Chemical composition of soils.' In *Chemistry of the Soil* ed. F.E. Bear. New York: Reinhold

Jackson, R. 1972. 'A vicious circle? – The consequences of von Thunen in tropical Africa.' *Area* 4: 258-61

Jacobs, H.S., R.M. Reed, S.J. Thien, and L.V. Withee, eds. 1971. *Soils Laboratory Exercise Source Book.* Madison: American Society of Agronomy

Jacoby, E.H. 1953. *Inter-relationship between Agrarian Reform and Agricultural Development.* Agricultural Study No. 26. Rome: FAO

James, R.W. 1971. *Land Tenure and Policy in Tanzania.* Nairobi: East African Literature Bureau

Johnson, R.A. and A.E. Tiarks. 1969. *Land Resources of South-east Tanzania.* Dar es Salaam: FAO (mimeo)

Johnson, W.M. 1963. 'The pedon and the polypedon.' *Proceedings, Soil Science Society of America* 27: 212-15

Johnston, B.F. 1970. 'Agriculture and structural transformation in developing countries: A survey of research.' *Journal of Economic Literature* 8: 369-404

Jones, T.A. 1959. 'Soil classification – a destructive criticism.' *Journal of Soil Science* 10: 196-200

Kardos, L.T. 1964. 'Soil fixation of plant nutrients.' In *Chemistry of the Soil*, ed. F.E. Bear. New York: Reinhold

Kellogg, C.E. 1963. 'Why a new system of soil classification?' *Soil Science* 96: 1-5

Kesseba, A., J.R. Pitblado, and A.P. Uriyo. 1971. 'The use of the 7th Approximation in Tanzania: A case study.' *Proceedings, Second Conference on Land Use in Tanzania.* Morogoro: University of Dar es Salaam. (mimeo)

– 1972. 'Trends of soil classification in Tanzania: The experimental use of the 7th Approximation.' *Journal of Soil Science* 23: 235-47

Kesseba, A. and A.P. Uriyo. 1971. 'Response of maize to NPK fertilizer application in Morogoro.' *Proceedings, Second Conference on Land Use in Tanzania.* Morogoro: University of Dar es Salaam (mimeo)

Kimambo, I.N. and A.J. Temu, eds. 1969. *A History of Tanzania.* Nairobi: East African Publishing House

Klingebiel, A.A. 1958. 'Soil survey interpretation – capability groupings.' *Proceedings, Soil Science Society of America* 22: 160-3

Kubiena, W.L. 1958. 'The classification of soils.' *Journal of Soil Science* 9: 9-19

Leeper, G.W. 1956. 'The classification of soils.' *Journal of Soil Science* 7: 59-64

Le Mare, P.H. 1959. 'Soil fertility studies in three areas of Tanganyika.' *The Empire Journal of Experimental Agriculture* 27: 197-222

Lewis-Bernard, J. 1961. *A Digest of Uluguru Customary Law.* Dar es Salaam: Faculty of Law, University of East Africa (mimeo)

Lock, G.W. 1969. *Sisal: Thirty Years' Sisal Research in Tanzania*, 2nd ed. London: Longman

Logan, M.I. 1972. 'The development in the less developed countries.' *Australian Geographer* 12: 124-53

Loxley, J. 1972. 'Tanzania – Economy.' In *Africa South of the Sahara, 1972.* London: Europa Publications, Ltd

Loxton, R.F. n.d. *A Simplified Soil Survey Procedure for Farm Planning.* Union of South Africa: Soil Research Institute, Department of Agricultural Technical Services

– 1961. 'A modified chart for the determination of basic soil textural classes in terms of the International Classification for soil separates.' *South African Journal of Agricultural Science* 4: 507-12

Mabbutt, J.A. 1968. 'Review of concepts of land classification.' In *Land Evaluation*, ed. G.A. Stewart. Melbourne: Macmillan

Macfarlane, J.S. 1968. *Livestock on Sisal Estates*. Tanga (Tanzania): Livestock Breeding Station (mimeo)

Macvicar, C.N. 1969. 'A basis for the classification of soil.' *Journal of Soil Science* 20: 141-52

Maignien, R. 1966. *Review of Research on Laterites*. Paris: UNESCO, Natural Resources Research No. IV

Maini, K.M. 1967. *Land Law in East Africa*. London: Oxford University Press

McAuslan, J.P.W.B. 1967. 'Control of land and agricultural development in Kenya and Tanzania.' In *East African Law and Social Change*, ed. G.F.A. Sawyers. Nairobi: East African Publishing House

McKay, J. 1968. 'A review of rural settlement studies for Tanzania.' *East African Geographical Review* 6: 37-49

McKay, J., A. Daraja, and W. Mlay. 1970. Interim Report on a *Base-line Study of the Proposed Village Settlement at Kiwanda*. Dar es Salaam: BRALUP, Research Report No. 2

Mellor, J.W. 1966. *The Economics of Agricultural Development*. Ithaca: Cornell University Press

Mifsud, F.M. 1967. *Customary Land Law in Africa*. Rome: FAO, Legislative Series No. 7

Mikenberg, N., G.R. Suggett, and J.W. Dewis, 1968. *Survey and Plan for Irrigation Development in the Pangani and Wami River Basins: Soils*. Rome: FAO

Millikan, M.F. and D. Hapgood. 1967. *No Easy Harvest: The Dilemma of Agriculture in Underdeveloped Countries*. Boston: Little, Brown, and Co.

Milne, G. 1935. 'Some suggested units of classification and mapping, particularly for East African soils.' *Soil Research* 4: 183-98

- 1936. 'A provisional soil map of East Africa.' *Amani Memoirs*. No. 38. Amana (Tanzania): Government Printer

- 1947. 'A soil reconnaissance journey through parts of Tanganyika Territory, December 1935 to February 1936.' *Journal of Ecology* 35: 192-265

Moore, J.E. 1971. *Rural Population Carrying Capacities of the Districts of Tanzania*. Dar es Salaam: BRALUP, Research Paper No. 18

Muir, J.W. 1969. 'A natural system of soil classification.' *Journal of Soil Science* 20: 153-66

Munsell. 1954. *Munsell Soil Color Charts*. Baltimore: Munsell Color Co.

Murdock, C. and J.P. Andriesse. 1964. *A Soil and Irrigability Survey of the Lower Usutu Basin (South) in the Swaziland Lowveld*. London: MMSO

Nasser, S.F. 1965. 'Statutory and customary land tenure.' In *Agricultural Development in Tanzania*, ed. H.E. Smith. Dar es Salaam: Institute of Public Administration

Newiger, N. 1968. 'Village settlement schemes: The problems of co-operative farming.' In *Smallholder Farming and Smallholder Development in Tanzania*, ed. H. Ruthenberg. Munich: Weltforum Verlag

Nieuwolt, S. 1973. *Rainfall and Evaporation in Tanzania*. Dar es Salaam: BRALUP, Research Paper No. 24

Northcote, R.C. 1945. *A Memorandum on Native Land Tenure*. Dar es Salaam: Government Printer

Ntemo, F.D. 1956. 'Some notes on Ngulu.' *Tanganyika Notes and Records* 45: 15-19

Nyerere, J.K. 1958. 'Mali ya Taifa/National property.' In *Freedom and Unity*, J.K. Nyerere. Dar es Salaam: Oxford University Press

- 1962a. 'President's Inaugural Address.' In *Freedom and Unity*, J.K. Nyerere. Dar es Salaam: Oxford University Press

- 1962b. 'Ujamaa – The basis of African socialism.' In *Freedom and Unity*, J.K. Nyerere. Dar es Salaam: Oxford University Press

- 1967. *Ujamaa Vijijini: Socialism and Rural Development*. Dar es Salaam: Government Printer

Obeng, H.B. 1968. 'Land capability classification of the soils of Ghana under practices of mechanical and hand cultivation for crop and livestock production.' *Transactions, Ninth International Soil Science Congress* 4: 215-23

O'Connor, A.M. 1967. *An Economic Geography of East Africa*. London: Bell

Oldaker, A.A. 1957. 'Tribal customary land tenure in Tanganyika.' *Tanganyika Notes and Records* 47/48: 117-44

Osborne, J.F. 1970. *An Appraisal of Mainland Tanzania in Respect of Rainfall, Potential Evaporation and Potential Land Use in Terms of River and Lake Catchments*. Dar es Salaam: Ministry of Agriculture, Food and Co-operatives, Research Notes on Soils and Water No. 1 (mimeo)

Page, J.B. and L.D. Baver. 1940. 'Ion size in relation to fixation of cations by colloidal clay.' *Proceedings, Soil Science Society of America* 4: 150-5

Pallister, J.W. 1971. 'The tectonics of East Africa.' In *Tectonics of Africa*. Paris: UNESCO, Earth Science Series No. 6

Papadakis, J. 1969. *Soils of the World*. Amsterdam: Elsevier

Peech, M. 1965. 'Hydrogen ion activity.' In *Methods of Soil Analysis*, ed. C.A. Black et al. Madison: American Society of Agronomy, Monograph No. 9

Penman, H.L. 1948. 'Natural evaporation from open water, bare soil, and grass.' *Proceedings, Royal Society (London)*, Series A, 193: 120-45

Pitblado, J.R. 1970. *A Review of Agricultural Land Use and Land Tenure in Tanzania*. Dar es Salaam: BRALUP, Research Notes No. 7

- 1975. 'Land capability and land tenure: Problems and prospects for agricultural development in the North Mkata Plain, Tanzania.' Unpublished doctoral dissertation. Toronto: Department of Geography, University of Toronto

Porter, P.W. 1973. 'Review of *The Geography of Modernization in Kenya: A Spatial Analysis of Social, Economic, and Political Change*, by E.W. Soja, 1968.' *Geographical Analysis* 5: 67-73

Pratt, D.J., P.J. Greenway, and M.D. Gwynne. 1966. 'A classification of East African rangeland.' *Journal of Applied Ecology* 3: 369-82

Quennall, A.M., A.C.M. McKinlay, and W.G. Aitken. 1956. *Summary of the Geology of Tanganyika.* Dodoma: Geological Survey of Tanganyika, Memoir No. 1

Rald, J. 1970. 'Ujamaa — Problems of implementation.' *Proceedings, East African Agricultural Economics Society Conference.* Dar es Salaam: University of Dar es Salaam (mimeo)

Rapp, A., V. Axelsson, L. Berry, and D.H. Murray-Rust. 1972. 'Soil erosion and sediment transport in the Morogoro River Catchment, Tanzania.' *Geografiska Annaler* 54 (A):125-55

Rattray, J.M. 1960. *The Grass Cover of Africa.* Rome: FAO. Agricultural Study No. 49

Raup, P. 1963. 'The contribution of land reforms to agricultural development: An analytical framework.' *Economic Development and Cultural Change* 12: 1-21

Richards, L.A., ed. 1954. *Diagnosis and Improvement of Saline and Alkali Soils.* Washington: United States Department of Agriculture, Agricultural Handbook No. 60

Rijkebusch, J. 1967. *Notes on Leguminous Cover Crops in Sisal.* Mlingano (Tanzania): Tanganyika Sisal Growers' Association, Bulletin No. 44 (mimeo)

Russell, M.B., ed. 1959. *Water and Its Relation to Soils and Crops.* New York: Academic Press

Ruthenberg, H. 1964. *Agricultural Development in Tanganyika.* Munich: Springer Verlag

– ed. 1968a. *Smallholder Farming and Smallholder Development in Tanzania.* Munich: Welforum Verlag

– 1968b. 'Some characteristics of smallholder farming in Tanzania.' In *Smallholder Farming and Smallholder Development in Tanzania*, ed. H. Ruthenberg. Munich: Weltforum Verlag

Saggerson, E.P. 1969. 'Geology.' In *East Africa: Its Peoples and Resources*, ed. W.T.W. Morgan. Nairobi: Oxford University Press

Samki, J.K. 1970. *A Review of Past, Present and Proposals for Future Research on Tanzanian Soils.* Dar es Salaam: Ministry of Agriculture, Food and Co-operatives (mimeo)

Sandberg, A. 1974. *Socio-economic Survey of the Lower Rufiji Flood Plain, Part*

I: Rufiji Delta Agricultural Systems. Dar es Salaam: BRALUP, Research Paper No. 34

Saunders, W.M.H. and E.G. Williams. 1955. 'Observations on the determination of total organic phosphorus in soils.' *Journal of Soil Science* 6: 254-67

Scott, R.M. 1969. 'Soils.' In *East Africa: Its Peoples and Resources*, ed. W.T.W. Morgan. Nairobi: Oxford University Press

Semb, G. and P.K. Garberg. 1969. 'Some effects of planting date and nitrogen fertilizer in maize.' *East African Agricultural and Forestry Journal* 34: 371-80

Sharma, B.N. 1957. 'Preliminary report on Magole irrigation scheme.' Dar es Salaam: Water Development and Irrigation Division (typescript)

Simonson, R.W. 1971. 'Soil association maps and proposed nomenclature.' *Proceedings, Soil Science Society of America* 35: 959-65

Smith, H.E. ed. 1956. *Agricultural Development in Tanzania.* Dar es Salaam: Institute of Public Administration

Soil Survey Staff. 1951. *Soil Survey Manual.* Washington: United States Department of Agriculture, Agriculture Handbook No. 18

- 1960. *Soil Classification, A Comprehensive System: 7th Approximation.* Washington: United States Department of Agriculture

- 1964. *Supplement to Soil Classification System (7th Approximation).* Washington: United States Department of Agriculture

- 1967. *Supplement to Soil Classification System (7th Approximation).* Washington: United States Department of Agriculture

- 1975. *Soil Taxonomy.* Washington: United States Department of Agriculture, Agriculture Handbook No. 436

Sorrenson, M.P.K. 1967. *Land Reform in Kikuyu Country.* Nairobi: Oxford University Press

Tanganyika. 1957. *African Census Report.* Dar es Salaam: Government Printer

- 1958. *Review of Land Tenure Policy.* Dar es Salaam: Government Printer

- 1962. *Land Tenure Reform Proposals.* Dar es Salaam: Government Printer

- 1964. *Tanganyika Five-Year Plan for Economic and Social Development, 1st July, 1964 – 30th June, 1969.* Dar es Salaam: Government Printer

Tanzania. 1967a. *Atlas of Tanzania.* Dar es Salaam: Government Printer

- 1967b. *The Arusha Declaration and TANU's Policy on Socialism and Self-reliance.* Dar es Salaam: Government Printer

- 1969a. *Tanzania Second Five-Year Plan for Economic and Social Development, 1st July, 1969 – 30th June, 1974.* Dar es Salaam: Government Printer

- 1969b. *1967 Population Census. Vol. 1. Statistics for Enumeration Areas.* Dar es Salaam: Government Printer

Teale, E.O. and E. Harvey. 1933. 'A physiographic map of Tanganyika Territory.' *Geographical Review* 23: 402-13

Temple, P.H. and A. Rapp. 1972. 'Landslides in the Mgeta area, western Uluguru Mountains, Tanzania.' *Geografiska Annaler* 54(A): 157-93

Temu, P. 1973. 'The ujamaa experiment.' *CERES* 6: 71-5

Thomas, I. 1971. Tribes.' In *Tanzania in Maps*, ed. L. Berry. London: University of London Press

Thomas, M.F. and G.W. Whittington, eds. 1969. *Environment and Land Use in Africa*. London: Methuen

Thomas, R.G. and V. Vincent. 1959. 'Classification of soils for land use purposes in the Rhodesias.' *Proceedings, Third Inter-African Soils Conference* 1: 351-60

Thornthwaite, C.W. 1948. 'An approach toward a rational classification of climate.' *Geographical Review* 38: 55-94

Troll, C. 1963. 'Landscape ecology and land development with special reference to the tropics.' *Journal of Tropical Geography* 17: 1-11

Uchendu, V.C. 1967. 'Some issues in African land tenure.' *Tropical Agriculture* 44: 91-101

Udo, R.K. 1966. 'Transformation of rural settlement in British tropical Africa.' *Nigerian Geographical Journal* 9: 129-44

Uriyo, A.P. 1966. 'Some aspects of livestock production with special reference to feeding in the Morogoro District of Tanzania.' Kampala: University of East Africa (typescript)

- 1970. 'A critique of some soil classification maps in Tanzania.' *Proceedings, First Conference on Land Use in Tanzania*. Morogoro: University of Dar es Salaam (mimeo)

Van der Eyk, J.J., C.N. Macvicar, and J.M. de Villiers. 1969. *Soils of the Tugela Basin*. Natal: Town and Regional Planning Commission

Van Wambeke, A. 1967. 'Recent developments in the classification of the soils of the tropics.' *Soil Science* 104: 309-13

Vesey-Fitzgerald, D.F. 1963. 'Central African grasslands.' *Journal of Ecology* 51: 243-73

Webster, C.C. and P.N. Wilson. 1966. *Agriculture in the Tropics*. London: Longmans

Webster, R. 1968. 'Fundamental objections to the 7th Approximation.' *Journal of Soil Science* 19: 354-66

Wilbanks, T.J. and R. Symanski. 1968. 'What is systems analysis?' *Professional Geographer* 20: 81-5

Yonge, D.D. 1965. 'Land tenure reform.' In *Agricultural Development in Tanzania*, ed. H.E. Smith. Dar es Salaam: Institute of Public Administration

Young, A. 1973. 'Soil survey procedures in land development planning.' *Geographical Journal* 139: 53-64

Zimmerman, J.D. 1966. *Irrigation*. New York: Wiley